SERIES ON
Water: Emerging Issues and Innovative Responses

Lake Governance

Editors

Velma I. Grover

Adjunct Professor
Faculty of Environmental Studies
York University
Toronto, ON
Canada

Gail Krantzberg

Professor
Centre for Engineered and Public Policy
School of Engineering Practice
McMaster University
Hamilton, ON
Canada

CRC Press is an imprint of the
Taylor & Francis Group, an **informa** business

A SCIENCE PUBLISHERS BOOK

Cover illustration reproduced by kind courtesy of Prof. Gail Krantzberg (co-editor)

CRC Press
Taylor & Francis Group
6000 Broken Sound Parkway NW, Suite 300
Boca Raton, FL 33487-2742

© 2018 by Taylor & Francis Group, LLC
CRC Press is an imprint of Taylor & Francis Group, an Informa business

No claim to original U.S. Government works

Printed on acid-free paper
Version Date: 20180426

International Standard Book Number-13: 978-1-138-63375-9 (Hardback)

This book contains information obtained from authentic and highly regarded sources. Reasonable efforts have been made to publish reliable data and information, but the author and publisher cannot assume responsibility for the validity of all materials or the consequences of their use. The authors and publishers have attempted to trace the copyright holders of all material reproduced in this publication and apologize to copyright holders if permission to publish in this form has not been obtained. If any copyright material has not been acknowledged please write and let us know so we may rectify in any future reprint.

Except as permitted under U.S. Copyright Law, no part of this book may be reprinted, reproduced, transmitted, or utilized in any form by any electronic, mechanical, or other means, now known or hereafter invented, including photocopying, microfilming, and recording, or in any information storage or retrieval system, without written permission from the publishers.

For permission to photocopy or use material electronically from this work, please access www.copyright.com (http://www.copyright.com/) or contact the Copyright Clearance Center, Inc. (CCC), 222 Rosewood Drive, Danvers, MA 01923, 978-750-8400. CCC is a not-for-profit organization that provides licenses and registration for a variety of users. For organizations that have been granted a photocopy license by the CCC, a separate system of payment has been arranged.

Trademark Notice: Product or corporate names may be trademarks or registered trademarks, and are used only for identification and explanation without intent to infringe.

Library of Congress Cataloging-in-Publication Data

Names: Grover, Velma I., editor.
Title: Lake governance / editors, Velma I. Grover, Adjunct Professor, Faculty
  of Environmental Studies, York University, Toronto, ON, Canada, Gail
  Krantzberg, Professor, Centre for Engineering and Public Policy, School of
  Engineering Practice, McMaster University, Hamilton, ON, Canada.
Description: Boca Raton : CRC Press, [2018] | Series: Water: emerging issues
  and innovative responses | "A science publishers book." | Includes
  bibliographical references and index.
Identifiers: LCCN 2018014046 | ISBN 9781138633759 (hardback)
Subjects: LCSH: Water-supply--Management. | Water resources--Management. |
  Lakes--Management. | Boundaries--Economic aspects.
Classification: LCC HD1691 .L33 2018 | DDC 333.91/63--dc23
LC record available at https://lccn.loc.gov/2018014046

Visit the Taylor & Francis Web site at
http://www.taylorandfrancis.com

and the CRC Press Web site at
http://www.crcpress.com

Preface to the Series

Water is the lifeline for all life on the planet earth and is linked to the development of all societies and cultures. But this development of all societies and cultures comes with a tag for increased water demand: because of increasing population as well as because of increasing demand from competing sectors such as agriculture, industry, domestic consumption, and recreation. This has impacted both the water quality and water quantity. The planet is at a point where anthropogenic changes have started to interfere with natural processes, including water cycle, and ecosystem services. More innovative approaches will be required to deal with emerging problems and transforming challenges into opportunities. This involves finding innovative solutions of treating water pollution, better governance structure and other institutional framework, good legal structure, in addition to increased investment in water sector.

Water is at the core of sustainable development hence sustainable water management and governance and is vital for humanity and sustainable future of human society (including human health, environmental health, poverty reduction and economic growth). This water series, "Emerging Issues and Innovative Responses" is an attempt towards coming up with a comprehensive series which will look at various aspects of water science, management, governance, law, economics and its relation to (sustainable) development (including social and cultural attitudes), impact of climate change on water resources, including water, energy and food nexus.

Velma I. Grover

Preface

Having worked on water governance, especially as it relates to transboundary lakes, both the editors felt more literature addresses transboundary Rivers and River-commissions with very little academic attention to transboundary lakes. This book on Transboundary Lake Governance is an attempt to fill that gap. The book focuses on comparative analysis of governance structures by examining policy, legal and institutional structures of current transboundary commissions to develop a better understanding of good governance of transboundary lakes. Cooperation among nations sharing natural resources is important for sustainable use of the shared resources. Lakes contribute a large part to GDP in most of the countries and in some cases are also responsible for providing fisheries (for food, source of protein and livelihood). Climate change and associated risks and uncertainties add more complexity to the issues. This book explores current water governance challenges, knowledge gaps and recommends a framework for good lake governance.

In this book, the editors made an attempt to explore certain common themes in different transboundary lake settings to develop a general lake governance framework model. Authors were invited to address the issues below in their respective chapters to bring out comparative analysis to develop a common transboundary lake governance framework.

- ❖ What are the critical/emerging science issues leading to policy change or formation of basin organizations?
- ❖ What is/was the driver for forming Commissions?
- ❖ Is there a legally binding treaty or is it soft law?
- ❖ Are any indicators of good governance, effective governance used?
- ❖ Is there integration of social/cultural and economic issues with ecological/environmental issues? Can these be described?
- ❖ How is science used in transboundary or adaptive transboundary governance?

Although when the book was perceived and started there was a hope to engage a lot of international professionals in transboundary lake studies globally, the final result does not have as many case studies that we expected. Still we believe the various case studies offer a rich literature on different transboundary lakes and addresses the academic literature gap in this area.

Velma I. Grover
Gail Krantzberg

Contents

Preface to the Series   iii
Preface   v

1. Transboundary Lake Governance   1
   Velma I. Grover and *Gail Krantzberg*

Section 1: Introduction

2. Water Governance Indicators: Challenges and Prospects for Improving Transboundary Lake Governance   11
   Carolyn Johns

3. Groundwater Governance and Assessment in a Transboundary Setting   40
   Sharon B. Megdal and *Jacob D. Petersen-Perlman*

Section 2: Case Studies

4. Transboundary Water Governance in the Great Lakes Region   67
   Victoria Pebbles

5. Legal Aspects of Transboundary Lakes Governance in the Western Balkans   78
   Slavko Bogdanovic

6. Governance of Transboundary Water Commissions: Comparison of Operationalizing the Ecosystem Approach in the North American Great Lakes and the Baltic Sea   111
   Savitri Jetoo-Åbo and *Marko Joas-Åbo*

7. Learning from the Transboundary Governance of Lake Victoria's Fisheries   130
   Ted J. Lawrence, James M. Njiru, Fiona Nunan, Kevin O. Obiero, Martin Van der Knaap and *Oliva C. Mkumbo*

8. Lake Titicaca: Case Study 151
   Paul Fericelli

9. Transboundary Governance in North America: 158
   More than 100 years of Development, Operation, and Evolution
   of the International Joint Commission
   Gail Krantzberg and *Velma I. Grover*

Index 173

CHAPTER 1

Transboundary Lake Governance

Velma I. Grover* and Gail Krantzberg

Introduction

Lakes play an important role in the economy and are critically important for human activities such as drinking, irrigation, fishing (for a lot of developing countries, fish is the largest source of protein), navigation, and even recreation. A lot of literature has been written on transboundary rivers, their treaties and commissions, little has been done on transboundary lakes (their conflicts and treaties including commissions). Globally, there are over 27 million natural lakes and about half a million artificial lakes, and the total volume of these lakes constitute about 70% of the available surface freshwater (Texas State University 2014). From these, there are about 1,600 transboundary lakes and 455 transboundary aquifers (Transboundary Waters Assessment Program, n.d.).

This book focuses mainly on transboundary lakes with some reference to transboundary groundwater as well. The book explores transboundary lake management and how treaties (or lack of) impact lake governance mechanisms and also the quality of water, because treaties play an important role in transboundary lake governance by establishing mechanisms for conflict resolution, increasing cooperation (Brochmann 2009, Wolf 1997) and in some cases also establishing guidelines for maintaining water quality (IJC, GLWQA). In general, transboundary lakes face numerous management and governance challenges because of factors such as: transboundary lakes do not recognize or respect political boundaries; they are under stress because of increased water pollution, overexploitation of resources, unsustainable water consumption coupled

---

1280 main st. w., hamilton, on, l8s 4k1, canada.
Email: krantz@mcmaster.ca
* Corresponding author: velmaigrover@yahoo.com

with low efficiency of water use, and poor implementation of management practices (where they exist) (Uitto 2002).

This chapter starts with a general introduction to transboundary lake governance, followed by a roadmap to the book, introducing various chapters of the book, concluding with results drawn from comparing findings in various chapters.

Transboundary Lake Governance

Although about 70% of earth is covered with surface water, about 97.5% in salt water and only 2.5% is freshwater (United States Geological Survey 2016). Out of this 2.5%, lakes contain about 67.5% of the total available surface freshwater on the planet, making lakes the largest source of surface freshwater (Intelligence Community Assessment 2012). Some of the earlier reports on transboundary lake governance, such as Lake Basin Management Initiative have looked at various lakes to make suggestions for improved lake governance at various levels such as local, basin, national or global levels (LakeNet, n.d.); the study by UNECE looks at the transboundary rivers, lakes and groundwaters in both Europe and Asia and provides an analysis on the status of water quality and quantity including pressure on water supply, transboundary impacts, etc. (Lipponen 2011). The convention on the Non-navigation of International Water Sources does provide guidelines and framework for equitable and reasonable water use (including prohibition to cause significant harm) and a conflict resolution clause (Mohamoda 2003) but it is vague in terms of providing guidelines for water allocations and enforceable international guidelines for transboundary lakes (Beaumont 2000). Basically, the convention does not have legal tools to implement sustainable water use and to ensure access water to individuals.

Transboundary lakes are the water bodies that cross political/national boundaries. Effective management of these lakes thus requires cooperation of the riparian countries involved and hopefully harmonization of water management policies along the lake. This may take a form of a treaty which can be a legally binding or non-binding instrument among one or more riparian states. Alternatively, institutions, such as commissions or organizations, can be established to deal with either the whole transboundary basin or specific issues such as fisheries or water quality. In general, lake management and governance needs to be rooted at the nexus of science and policy, where science needs to go beyond just open waters to include the wider watershed and socio-economic issues that will impact water quality and quantity. As joint commissions are established to institutionalize cooperation mechanisms and conflict resolution processes, it is important that institutional arrangements also expand to support development of these linkages between science (scientists) and policy (policy/decision makers) which will hopefully foster integrated water resources management. Successful transboundary lake management requires open communication for information exchange and data sharing to establish joint monitoring programs among nations and to explore benefit sharing among riparian countries for the entire basin.

A study (Lubner 2015) has shown that between 1990 and 2013, 52 international water treaties related to large lakes were implemented and out of this 24 focused mainly on joint management but only 4% of these treaties included enforcement mechanism. Other treaties focused mainly on water quality and quantity while none focused on border issues or ground water or recreation. Most of these treaties were signed during the 1990s (39) while the rest of the 13 treaties were implemented during the 2000s. Also, between 1990 and 2013, 53 international water conflicts were registered for the 17 of the 35 largest transboundary lakes but none of them was violent. Most of these conflicts were reported in Africa and Lake Malawi alone had 15 conflicts between Malawi and Tanzania. Out of these 53 reported conflicts, 20 were because of water quantity (mainly around Aral Sea and Lake Victoria), 14 were due to border issues (between Malawi and Tanzania). Although most of the treaties were signed in the 1990s, more conflicts (40 of the 53) were reported in 2000s. The study also concludes that having a treaty among riparian nations does not always influence conflict or cooperation, there are some external factors that will impact conflicts. These external factors include: location, political regime, population, and environmental variables; water scarcity; water mismanagement and governance (since most of the water conflicts relate to water allocation and water use) (Lubner 2015). The next sections discuss some of these elements as described by the authors of this book.

Roadmap to the Book

In the first section of the book, the first chapter by Jones, discusses terms such as "good governance" and "transboundary water governance" in general, good governance as it relates to water management and some Water Governance Indicators/Indexes (WGIs). This is followed by a discussion on water governance indicators and importance of water indicators to understand complex water governance systems; measure effectiveness of water policies and performance of water governance institutions; improve policy outcomes; and increase accountability in water governance regimes. The chapter then dives into the common WGIs used to assess progress in water governance systems and the applicability of those indicators in transboundary lake systems. As can be seen from some of the case studies in this book, transboundary water governance is complex and better approaches are needed to fully understand and manage these complex transboundary water systems. Water governance indicators in a transboundary water governance setting seems like a good approach for monitoring of transboundary systems. But as the author has raised the question—these indicators were mainly developed within countries and not really for transboundary systems, would these be really transferable to a transboundary setting? The chapter concludes with reflections on the potential and challenges of using water governance indicators to improve transboundary lake governance and water governance in the future.

In the next chapter Megdal and Petersen-Perlman discuss increasing reliance on groundwater on the planet. Since water demand is increasing due to increased global population, industrialized agriculture and industries reliance on groundwater has increased. Groundwater has been overused because of latest inventions such as centrifugal pump and deep well pumping technology. Unlike surface water, groundwater is invisible hence it is difficult to quantify reductions of water in storage, or the increased depth of drilling, it is hard to detect pollutants (and even harder to remedy). This chapter deals with a complicated theme of transboundary groundwater management, although groundwater is part of the transboundary surface water system, transboundary surface and groundwater systems are often governed and managed in separate regimes. Also, the hydrologic relationships between surface water systems and groundwater systems are often not very well-understood which makes it even harder to determine the basic sovereignty of groundwater resources. The authors provide an overview of groundwater governance in practice followed by a discussion on groundwater governance in a transboundary setting, including consideration of commonly accepted principles for governing groundwater and recent developments in legal principles for transboundary groundwater governance. The authors then consider a case study of groundwater assessment along the United States (U.S.)—Mexico border in the context of these principles and explain the value of the cooperative guiding framework established for this assessment.

Pebbles in "**Transboundary Water Governance in the Great Lakes Region**" examines institutional and policy frameworks that govern transboundary water management in the Great Lakes. The author looks at the historic and current institutions along with policy arrangements with a discussion on water management successes and opportunities in the Great Lakes region with a reflection on gaps and opportunities for future water management options in the Great Lakes regions as well as other large bilateral and multilateral aquatic ecosystems.

In **Legal Aspects of Transboundary Lakes Governance in the Western Balkans,** Bogdanovic discusses current legal aspects of the governance systems applied or under development in the catchment areas of several natural freshwater lakes in the southern part of the Balkan Peninsula. The lake systems discussed in the chapter include: Skadar/Shkodra (shared by Albania and Montenegro), Ohrid (shared by Albania and Macedonia), Macro and Micro Prespa (shared by Albania, Greece and Macedonia) and Dojran (shared by Macedonia and Greece) lakes. The author discusses some legal aspects of "governance" followed with a concise review of the most important legal instruments applicable at the level of the lake basins in review as well as description of the governance concepts identified. The author concludes the chapter with some legal/institutional elements of the systems of lake governance that could be identified as features of a desirable "good" transboundary waters governance practice.

In the next chapter, Jettoo and Joas compare The Laurentian Great Lakes in North America and the Baltic Sea in Europe. The authors point out some similarities in both cases such as: signing of transboundary governance agreement in the early 1970s due to concerns of water pollution. The Great Lakes Water Quality Agreement (GLWQA) was signed between Canada and the United States of America (US) amidst concerns about the 'dying' Lake Erie in 1972. At the same time in Europe, negotiations were ongoing for a transboundary agreement amidst concerns about increasing pollution to the Baltic Sea resulting in a transboundary water agreement in 1974, the Helsinki Convention, which established the Helsinki Commission (HELCOM) as the coordinating body. The International Joint Commission in the North American was established in 1909 much before the GLWQA was signed. Whilst these commissions can be seen as successes because they were effective in bringing the key national players to the table, the continued degradation of these transboundary water ecosystems would suggest that they are not yet successful in applying the ecosystem approach to governance, as called for in both transboundary agreements and deemed necessary to achieve the purposes of the agreements. This chapter also looks into the effectiveness of these transboundary water commissions in operationalizing the ecosystem based approach to governance by assessing their adaptive capacity, the governance capacity for dealing with change. The authors have used a framework for adaptive capacity to assess the performance of these transboundary commissions against these principles. The authors then identify gaps, obstacles (for example, the federal structure of US and Canada can be an obstacle on the North American side or geo-political position of Russia and lack of assessment of measures on the Baltic side are clear obstacles) and makes recommendations that can inform policymakers.

Lawrence et al. have used Lake Victoria's fisheries governance system as a case study to discuss management on this multi-national, freshwater resource. The authors have described some past influences which have limited successful management of the fisheries resources, the impetus for current management approaches, and the successes and ongoing challenges that currently exist. Lake Victoria, East Africa, is a transboundary resource, and is an important multi-use resource, known distinctly for its valuable, highly productive, and diverse fisheries. The lake employs three million people in fisheries-related activities, contributes USD 600 million annually to the governments of the riparian countries and provides food and livelihood security to 20 million people in the region. Since Lake Victoria is categorically a social-ecological system, it is important to establish institutions to enable a coordinated, harmonized approach to governance, with the intent to ensure sustainable use of the resources. The chapter discusses the two transboundary efforts on Lake Victoria namely Lake Victoria Basin Commission (which covers tourism, agriculture, industry and transportation) and Lake Victoria Fisheries Commission (with a focus mainly on fisheries, which adds to the economy, as well as needed for subsistence as a source of protein to deal with the challenges of over-harvesting).

The next case study on Lake Titicaca by Fericelli gives a brief description of drivers impacting transboundary challenges in Lake Titicaca such as environmental degradation, the legal framework and climatic conditions. This is followed by a discussion on how conflicts in managing area are resolved by Bolivia and Peru and their view on using dams in the region. Lastly, the author discusses how social/cultural and economic issues influence restoration of Lake Titicaca.

In the last chapter in the case study section Krantzberg and Grover discuss the historical development of water cooperation mechanisms such as creation of Boundary Waters Treaty (1909). A historical development of the International Joint Commission (IJC) and evolution of the Great Lakes Water Quality Agreement (GLWQA) followed by latest developments in the GLWQA to address more current water challenges. As discussed in the chapter, the Boundary Waters Treaty was one of the visionary and pioneering treaties ahead of its time, introducing the concept of transboundary governance using an ecosystem approach for some current challenges including the Devil's lake incident (and in general lesser number of references to IJC) and its failure to reproduce results for algal blooms in Lake Erie suggest that it is time to update some other components of the treaty in addition to updating GLWQA.

Conclusion

Based on the chapters in the book, some common themes emerged. For example, most of the drivers impacting transboundary problems (leading to cooperation) include:

Impact of growing population and development around lakes leading to issues of increased urbanization, deforestation, increased demand for water from growing cities, industrialization and intense agriculture (to feed the growing population). This has often resulted in both water quality (toxic chemicals from both industry and agriculture) and quantity problems, and water allocation issues—often bringing riparian countries sharing the water body to the table for discussions, cooperation and good governance. Both academicians and policy makers realize that with increasing population, global water demand is expected to increase by 55% by 2050 and resulting water crises will be governance crises, hence there is a move towards good governance for water resources within countries as well as in shared transboundary systems (Jones, Chapter 2).

Invasive species once introduced to the system are not easy to eradicate or even control—a basin wide effort to monitor these species is needed in addition to an institutional system to address this problem. Since invasive systems can cause both economic impacts (for example it has been bad for North American economy but a driver in Lake Victoria) and ecological damage impacting human health, a plan to manage them would need engagement from all the riparian countries.

Climate change is expected to influence the water levels in the lake (most likely impact will be decrease in water levels) and the increasing population will also put pressure on water demand from lakes which can be a cause of potential conflict (Wilner 2005). Climate change is not just affecting chemical, biological, and physical aspects of the North American Great Lakes but is also leading to fluctuating water levels over the years—from really low around 2012/13 to an extreme high in 2016 (as the climate change happens, amount of precipitation varies). For most of the lakes, the common issue is increased temperatures and impacts related to this warming (impacting nutrient and bio-geochemical cycles).

Formation of institutions/treaties that are established and signed among riparian nations are responsive in nature. All the case studies in the book reflect this. Even in North America, the Boundary Waters Treaty of 1909 (that also established IJC) was signed in response to water quantity issues raised by hydropower development and water quality problems (causing human health concerns). Similarly, the Great Lakes Fishery Commission came into existence to respond to the Sea Lamprey crisis. In most cases, a binding legal treaty is lacking and for most of the treaties political/economic power asymmetry plays a role. Also, there is a lack of open communication and information flow among riparian countries in terms of data sharing (though the Lake Titicaca case study does write about a binational monitoring system). Although, a few various indicator metrics have been established (as discussed by Jones) their implementation in most of these transboundary basins is non-existent.

Stakeholder engagement is important. One of the best examples of the bottom up approach from all the examples in the book are the beach management units used in African Great Lakes which reflect connection to watershed and sub-watershed management with intense civic engagement with support from government and technical expertise.

Just like transboundary lakes, transboundary groundwater and aquifers have not received attention and have weak institutional structures but there is a growing literature and movement towards strengthening these institutions (Medgal and Petersen-Perlman 2018, see Chapter 3 of this book). Moving forward, efforts need to be put into building capacity of current institutions to deal with the current problems and increasing threats related to climate change.

References

Beaumont, P. (2000). The 1997 UN Convention on the law of non-navigational uses of international watercourses: its strengths and weaknesses from a water management perspective and the need for new workable guidelines. *International Journal of Water Resources Development*, 16(4): 475–495.

Brochmann, M.A. (2009). Peaceful management of international river claims. *International Negotiation*, 14(2): 393–418.

Intelligence Community Assessment. (2012). Global Water Security. Retrieved from http://www.dni.gov/files/documents/Special%20Report_ICA%20Global%20Wate.

LakeNet. (n.d.). World Lake Basin Managament Initiative. Retrieved from http://www.worldlakes.org/programs.asp?programid=2.

Lipponen, A.C. (2011). Second assessment of transboundary rivers, lakes and groundwaters. Economic Commission for Europe. Retrieved from http://www.unece.org/fileadmin/DAM/env/water/publications/assessment/English.

Lubner, V. (2015). The Impact of International Water Treaties on Transboundary Water Conflicts: A Study Focused on Large Transboundary Lakes. University of Wisconsin-Milwaukee.

Mohamoda, D. (2003). Nile basin cooperation (no. 0280-2171).

Texas State University. (2014). TWAP Transboundary Lakes Project: Lakes The Mirrors of the Earth. Retrieved from http://icws.meadowscenter.txstate.edu/TWAP.html.

Transboundary Waters Assessment Program. (n.d.). Retrieved from http://www.geftwap.org/twap-project.

Uitto, J.A. (2002). Management of transboundary water resources: lessons from international cooperation for conflict prevention. *The Geographical Journal*, 168(4): 365–378.

United States Geological Survey. (2016). The Water Cycle: Freshwater Storage. U.S. Department of the Interior and U.S. Geological Survey. Retrieved from http://water.usgs.gov/edu/watercyclefreshstorage.html.

Wilner, A. (2005). Freshwater scarcity and hydropolitical conflict: Between the science of freshwater and the politics of conflict. *Journal of Military and Strategic Studies*, 8(1).

Wolf, A. (1997). *Water and Human Security*.

Section 1
Introduction

CHAPTER 2

Water Governance Indicators
Challenges and Prospects for Improving Transboundary Lake Governance

Carolyn Johns

Introduction

Around the globe, scholars, practitioners and a number of international organizations have been increasingly focused on improving governance and policy outcomes related to a wide range of social, economic and environmental problems. Internationally, 'good governance' has become somewhat of a holy grail in both developed and developing countries. For the past two decades, scholars and practitioners at the international and domestic levels have been using the concept of 'good governance' as part of a quest for improving human governance of the world's ecosystems in a period of population growth, climate change, and increasing human demands on the natural environment. Good governance of water resources, within countries and in shared transboundary water systems, is part of this pursuit given predictions that global water demand will increase by 55% by 2050 (OECD 2015a) and that both scholars and policy practitioners realize that "water crises are primarily governance crises" (OECD 2015a). Water governance is also a critical part of the United Nations Water Action Decade 2018–2028 (UN 2018).

Department of Politics and Public Administration, Ryerson University, 350 Victoria Street, Toronto, Ontario, Canada, M5B 2K3.
  Email: cjohns@ryerson.ca

Governance is a concept that is used in theory and research in a wide range of scholarly and professional fields. At the heart of governance as a concept and governance theory related to the environment and water is a broad understanding that addressing complex environmental problems requires collective action at a variety of scales. Governance involves both state/government and non-state, non-government organizations and actors trying to manage human-environment problems at the global, transboundary, regional, national, subnational and local scales. Governance theory has influenced a wide range of fields and virtually every subfield in political science, in particular public policy and public administration (Frederickson et al. 2016). Governance has become a central concept and governance theory has been very influential in scholarship on environmental policy and water policy.

Scholars and practitioners from various disciplines are engaged in water governance and water policy research and the multi-level governance and transboundary challenges related to water governance are well documented. Transboundary water governance is also at the heart of many other global governance challenges. Concern exists in both developing and developed countries about both systems of abundance and systems of scarcity and the complex geo-political context in which transboundary lakes and rivers are governed. With growing global concern about water governance, particularly related to climate change, international organizations such as the United Nations (UN), through its Millennium Development Goals, and the Organization for Economic Cooperation and Development (OECD), through its water governance program, have been keenly interested in assessing and promoting better water governance and policy regimes. Water governance is now on the agendas of many international organizations and scholarly research teams.

Whether research in this area is classified as water governance research or water policy research, there are scholars from a wide range of disciplines and fields doing research 'for' policy, 'about' policy and 'on' policy, many with a focus on the outcomes of water governance systems or regimes. Although traditionally policy scholars focused on public policies developed within the context of the nation state and comparative analysis across nation states, the concept of governance has now cast the policy research net quite broadly as policy scholars, and scholars and practitioners from a wide range of fields accept the global and international dimensions of public policy and the wide range of actors and institutions from the public, private, and non-government sectors who are involved in collective action related to water governance. As outlined in other chapters in this book, scholarship and practice focused on transboundary lake and river system governance highlights the complex realities of governance and the quest for better approaches to understanding and managing complex transboundary water systems. A central question in this chapter is whether the kind of theoretical and methodological work on indicators, which have primarily been developed within countries or jurisdictions, is transferable to the transboundary context and if so, what are the challenges and prospects for using water governance indicators to improve transboundary lake governance.

This chapter introduces 'water governance' and 'transboundary water governance' as concepts that embody the increased faith and effort by scholars and practitioners to use water governance indicators to: understand complex water governance systems; measure effectiveness of water policies and performance of water governance institutions; improve policy outcomes; and increase accountability in water governance regimes. This chapter begins by briefly reviewing the scholarly and practitioner literature on good governance, water governance and transboundary water governance and outlines how various definitions of these concepts underpin water governance indicators. The second section defines Water Governance Indicators (WGIs) and situates these indicators in the broader literature on environmental indicators. This section also outlines how water governance indicators have become a point of integration for a number of different scholarly and practitioner fields that are grappling with complexity of water governance and transboundary systems. It also highlights the common WGIs used to assess progress in water governance systems and the applicability of those indicators in transboundary lake systems. The chapter concludes with reflections on the potential and challenges of using water governance indicators to improve transboundary lake governance and water governance in the future.

Water Governance and Transboundary Water Governance

Water governance is a concern across the globe. Both water quantity and water quality are critical issues on a global and national scale. Concerns about water governance are evident both within countries and for many countries which share transboundary water resources with neighboring jurisdictions. In both contexts Water Governance Indicators (WGIs) are being developed and used based on some faith that they can improve governance and policy outcomes. In order to assess the applicability of water governance indicators related to transboundary lake systems it is important to first define 'good governance', 'water governance' and 'transboundary water governance' as WGIs flow from the scholarly and practitioner definitions of these key concepts.

Many different international organizations and scholars from several different fields including environmental studies, geography, political science, economics, and international development studies, use the concept of good governance. The scholarly and practitioner literature on good governance is substantial and beyond the scope of this chapter. For purposes of this chapter the definition from the United Nations (UN) is a good starting point—governance is the process by which decisions are made and implemented (UN 2016). There is no scholarly or practitioner consensus on how exactly to define good governance and what exactly needs to be improved. However, there have been various attempts to define and quantify governance, particularly using an indicators approach (Muriithi et al. 2014). Definitions of 'water governance' share some foundations with definitions of governance and good governance. There is some consensus among scholars and practitioners that 'water governance' is a general concept related to improving and adapting governance regimes to meet current and future challenges. Some definitions broadly stem from different disciplinary foundations and from water ethics and values (Johns forthcoming) and others are more narrowly associated with very specific geographic and functional uses of water. Generally, all definitions include some aspect of collective authoritative action related to water as a natural and common resource.

How water governance is defined determines levels of analysis, units of analysis and the various research approaches and methodologies used to study water governance. Increasingly, the socio-ecological systems, adaptive systems and institutionally-grounded definitions from political science and policy studies are being combined resulting in interdisciplinary definitions of water governance. Pahl-Wostl et al. (2008) highlights that water governance is "the development and implementation of norms, principles, rules, incentives, informative tools, and infrastructure to promote a change in the behavior of actors at the global level in the area of water governance". Moss and Newig (2010) argue that dimensions of scalar politics are central to water resource management and there is a need for more integrated approaches. Scholars from this perspective have also used the concepts of 'governance capacity', and 'adaptive capacity', and 'institutional capacity' (Edelenbos and Teisman 2013, Hill and Engle 2013, Betinni et al. 2015) to integrate more interdisciplinary and dynamic dimensions into definitions of water governance.

Practitioners across the globe have also been focused on defining and improving water governance. The Global Water Partnership (GWP), a network with government and nongovernmental members from 182 countries from all continents but North America, seeks "a water secure world" through improving the sustainable governance and management of water resources. GWP defines water governance as "the range of political, social, economic and administrative systems that are in place to develop and manage water resources, and the delivery of water services, at different levels of society" (Rogers and Hall/GWP 2003, 7). In pursuit of sustainable water governance, GWP articulates three basic goals: (i) to catalyze change in policies and practice; (ii) to generate and communicate knowledge and (iii) to strengthen partnerships. It, like many other international organizations, strives to improve water governance through research and knowledge mobilization within and across developed and developing countries. Other international organizations define water governance as "the political, social, economic and administrative systems in place that influence water's use and management. Essentially, who gets what water, when and how, and who has the right to water and related services, and their benefits" (watergovernance.org 2016). "Governing water includes the formulation, establishment and implementation of water policies, legislation and institutions, and clarification of the roles and responsibilities of government, civil society and the private sector in relation water resources and services. The outcomes depend on how the stakeholders act in relation to the rules and roles that have been taken or assigned to them" (watergovernance.org 2016).

The OECD bases its definition of water governance on broader principles of good governance: legitimacy, transparency, accountability, human rights, rule of law and inclusiveness. As such, they consider water governance as a means to an end rather than an end in itself, and define water governance as, "the range of political, institutional and administrative rules, practices and processes (formal and informal) through which decisions are taken and implemented, stakeholders can articulate their interests and have their concerns considered, and decision-makers are held accountable for water management" (OECD 2015b). Based on a comprehensive inventory of 60 water governance programs, and work with a consortium of experts from member countries, the OECD has tried to advance its definition of water governance. In 2015, the OECD developed 12 water governance principles that have been adopted by all OECD member countries. Three key components are at the core of the 12 water governance principles: effectiveness, efficiency and engagement. These three components reflect the policy goals of most water policy regimes, both domestic and transboundary. As outlined in Fig. 2.1, the 12 water governance principles are the key factors that jurisdictions must have in order to achieve the inner circle goals and enhance water governance. These principles however are mainly conceptualized in the context of water governance within OECD member countries. In keeping with other OECD approaches to good governance and performance management, OECD member countries will then use these principles to report on the state of water governance within their borders. At the World Water Forum in 2018 the OECD released 36 WGIs, three for each of the 12 principles (OECD 2018).

16 *Lake Governance*

Figure 2.1. The OECD's Definition of Water Governance through Water Governance Principles.
Source: Organization for Economic Cooperation and Development (OECD) 2015b.

In addition to the scholarly and practitioner literature on water governance, there are also numerous definitions of 'transboundary water governance' (TWG). This concept is also used in a wide range of disciplines to capture the reality that water governance does not align with political boundaries of nation states or the political boundaries within nation states. It has been used to study and address problems that are transboundary from the watershed scale to the global scale. For purposes of this chapter the focus is on those who define transboundary water governance at the global scale involving shared waters across national boundaries.

Like water governance, there are scholars and practitioners with backgrounds in environmental studies, international relations, public policy, geography, science and engineering using the concept of transboundary water governance. The concept is thus interdisciplinary. Work by Earle et al. (2010) on transboundary water management falls into this category. Bressers and Lulof (2010), in keeping with more of a socio-ecological systems approach, argue that transboundary water governance requires cooperation through 'boundary spanning' strategies. They emphasize the important role of social capital, as well as administrative and political alliances, through cross-boundary collaboration. In addition, there are clusters of scholars who for some time have been interested in environmental regime effectiveness (Young 2001), water regime effectiveness and transboundary water regime effectiveness. Many of these scholars have their roots in international relations and focus on international agreements and 'soft law' as the foundation for transboundary water governance (Zeitoun et al. 2011). Others from this perspective call for less focus on international institutions and more focus on 'hybrid actors' in transboundary water governance research. Suhardiman and Giordano (2012) argue that the TWG research has been too-focused on state actors and institutions, stemming from those in international relations who emphasize research focused on conflict and cooperation in water systems. They are also critical of the focus on international agreements as outcomes of policy development as they don't focus attention enough on incapacity to govern and implement (Suhardiman and Giordano 2012).

In the past few years there has been increasing recognition by scholars and practitioners working on water governance that those focused on water governance within states face many of the same complexities as those focused on TWG. There is clear recognition of multilevel characteristics, the importance of scale, the complexities of water governance, and a deeply rooted commitment to improve governance and policy outcomes. When various disciplinary perspectives are brought to the fore, assessing water governance in hopes of improving outcomes is a tall order. Many of these scholars and practitioners also recognize that there is a serious challenge in trying to embrace complexity yet also research key factors or variables that can diagnose and explain outcomes. Water governance indicators work plays out at the nexus of these approaches to water governance and transboundary water governance.

In summary, in the past two decades there have been efforts on several disciplinary fronts in the academic literature and numerous efforts by practitioners internationally to define, understand and improve water governance. Several different fields and bodies of literature have been grappling with water governance and transboundary water governance as concepts at various scales. There is some evidence that while in the past research questions, level of analysis and key concepts were quite distinct between various fields in social science, there is increasing overlap in research foci, key concepts and the emphasis on regime effectiveness, governance outcomes and policy outcomes. As depicted in Fig. 2.2, the focus on water governance indicators (WGIs) can be situated at the interface of several fields of scholarship and practice.

18   Lake Governance

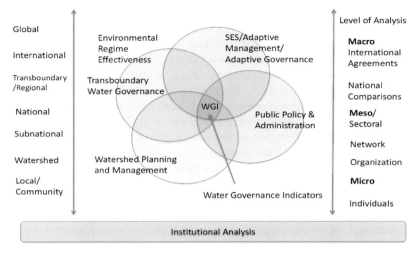

Figure 2.2. Water Governance Research and Water Governance Indicators (WGIs).

While this chapter does not provide the scope to comprehensively review the various bodies of scholarship that are focused on WGIs, a review of this literature indicates that most of the models focused on water governance and transboundary water governance incorporate a focus on scale and a focus on governance institutions as foundational in terms of understanding water governance systems and regimes.

If governance and policy regimes are to produce better outcomes in the next 40 years than have been achieved in the past 40 years, water governance and transboundary water governance need to be based on the recognition that complex problems require complex understandings of human and societal behavior and the reality that water systems are inherently multi-level, multi-scalar and transboundary. Many scholars and practitioners, are currently grappling with this complexity-measurement challenge. The next section provides a brief review of the evolution of water governance indicators before outlining several cases where WGIs are being applied at the transboundary scale, primarily related to river systems but also related to lake systems.

Governance Indicators and Water Governance Indicators

Very basically, indicators are quantitative and qualitative metrics designed to provide information on the state or condition of something and when tracked overtime to highlight progress or change. An indicator can be focused on a governance system or regime, parts of a governance system, or a policy process. Data for indicators can be comprised of one variable, a set of variables or an aggregation of variables in an index. Demand for indicators is increasing as jurisdictions as policy makers try to use best available evidence to determine and prioritize collective efforts and investments. Indicators are developed and used by a range of governance actors, advocates, scholars and practitioners. The aim of an indicator as a policy instrument is to provide information to policy and decision-makers and to assist them in management of a particular system (Lorenz et al. 2001). Indicator development involves the definition of what is being analyzed, construction of clear criteria based on theoretical conceptual links between problems and solutions, and clearly constructed methodologies based on these links. Indicators are used in a variety of contexts and sometimes called performance indicators, performance measures, effectiveness indicators and governance indicators.

There is now a voluminous literature on indicators. Over the past few decades, water governance indicators have generally evolved from broader environmental indicators. Environmental indicators evolved to provide information related to some baseline of the state of the environment, the extent and nature of environmental degradation, change in environmental components or performance of institutions and management tools (Bennet and Roche 2000). For example, the Environmental Sustainability Index a composite index published from 1999 to 2005 by Yale University's Center for Environmental Law and Policy, Columbia University's Center for International Earth Science Information Network (CIESIN), and the World Economic Forum. It tracked 21 elements of environmental sustainability, including 77 variables of which 10 related to water. Another example is the OECD's Core Environmental Indicators which were designed to track environmental progress and included five water-related response indicators, all related to the state of waste water treatment facilities in various countries. The UN and UNESCOs World Water Assessment Program (WWAP) was created in 2000 in response to "a well-recognized need to undertake comprehensive and objective assessments of all aspects of water resources, including the ability of nations to deal with water management challenges" (UN-Water 2008).

On the international scale many of these indicator initiatives aligned with Integrated Water Resource Management (IWRM) a concept and internationally recognized management assessment framework that helped to refine the scope of water policy assessments, and agreement on common principles to establish country comparisons (De Stefano 2010). Since 2003 the World Water Development Report has included water indicator data on a wide range of water-related issues. At the same time, there was growing criticism that environmental indicators were not using baseline ecosystem data and indicators based on narrow quantitative measures did not capture socio-ecological complexities (Dale and Beyeler

2001, Niemeijer 2002, Molle and Mollinga 2003). The OECD at the time was using two main WGIs: freshwater quality (waste water treatment connection rates) and freshwater resources (intensity of use). The OECD's water indicator work continued with the report *Managing Water for All* as a means of addressing chronic investment shortfalls for water and sanitation (OECD 2009). This report is also noteworthy for its explicit recognition of the governance capacity required for such instruments to be successful. A survey of 17 OECD countries in 2011, identified water governance gaps and called for a more 'systemic' approach to water policy to overcome critical multilevel governance challenges. A stated priority was that member states should be using "a multilevel approach integrating international, national and local actors can help diagnose inherent governance challenges" (OECD 2011) and that "further research should study 'micro-governance' to identify good local practices" (OECD 2011). To address this complexity the OECD developed a framework that focused on analyzing seven key implementation gaps related to water policy that member countries could collect data and report on.

By 2010, the international emphasis on 'good governance' and 'capacity-building' were well established and indicators of governance and good governance began to emerge in many policy domains. Water governance begins to replace IWRM as a central concept in water policy discourse and as the foundation for indicator development. At the same time, in the environmental and water policy area critiques continue about the inherent tension between indicators trying to simplify complex social and environmental phenomenon and the complex, multi-level, multi-sector, multi-actor reality of environmental problems. As Kaufmann and Kraay (2008) noted, "… any particular indicator of governance can usefully be interpreted as an imperfect proxy for some unobserved broad dimension of governance". The challenges, limitations and hazards of environmental indicators were by this time well recognized (Barnett et al. 2008). Nonetheless, in the past several years, indicators have become more sophisticated and indexes have gained in popularity. Efforts to streamline the number of indicators and use indexes have also been criticized by those who highlighted the problems associated with aggregating sets of quantitative measures in indices (Kaufman et al. 2005). Broader environmental indexes continue to include water governance indicators. The Yale Environmental Performance Index (EPI) for example evolved from having numerous indicators related to water to most recently having three key indicators related to water: access to drinking water; access to sanitation; percentage of population with water treatment/wastewater management. The Yale EPI measures these across 180 countries and argues they are the best indicators to date based on available comparative data.

The baseline target for these measures is a target of 100% of population (Yale EPI 2016).

Alongside the shift from a focus on program outputs, to program response, to outcomes and then to broader governance indicators, analysts have also become more open to qualitative indicators. Arguments for a more qualitative and context specific approach to indicator development came from scholars and practitioners who continued to reflect on the limitations of indicators. This becomes a central theme in the performance management and measurement literature in public administration as scholars focused on the challenges with measurement, the actual use of indicators, and the negative behavioral side-effects that performance indicators can have. Current governance indicators related to various environmental issues are now more context-specific, try to embrace scales and complexity and deploy a wider range of data collection and analytical methods. This same development path is evident related to water governance indicators. In addition, different water governance indicators have evolved for water security (Dunn and Bakker 2009, Norman et al. 2013, Garrick and Hall 2014); for water stress (TWAP 2016); and for water poverty (Sullivan 2002, Garriga and Foguet 2015).

In 2014–15, the OECD compiled an inventory of water governance indicators that revealed there has been considerable growth in the number of water indicators developed over time and the scope of their application. The OECD's inventory found that over 60 organizations were engaged in research related to assessment and performance indicators (OECD 2015b). This inventory included analytical frameworks and indicators for systems of scarcity and systems of abundance; developing and developed countries; water quality and quantity, lake and river systems. Most were focused on surface water, with some on groundwater. The inventory examined all of the indicator frameworks in terms of purpose, geographic coverage, indicators included, and those with specific indicators and data collection for water governance. Some were global/international, some regional and some national. Of those included in the OECD inventory some like Transparency International's Water Management Transparency Index focus on how much information about water resources is available to the public. Others, such as some of the scholarly models included governance related indicators such as the Equity Index for Water and Sanitation developed by Luh et al. (2013). Only 8 of the 60 indicator programs were classified as focusing on water governance indicators, however, on closer examination some of the other indicator suites included water governance indicators but did not label them as such.

Analyzing this inventory reveals that generations of water governance indicators are also evident, although only more recently labeled as such. Many early WGIs were part of broader suites of environmental indicators, then stand-alone water governance indicators emerged, primarily focused on within country measurement and cross-national comparisons. Like many frameworks and sets of performance indicators, the OECD's inventory of existing water governance indicators also reveals two general types: those that are data driven and those that are theory driven. Data-driven are those whereby data availability is the central criterion for indicator development and data is provided for all selected indicators. Theory-driven are those that focus on selecting the best possible indicators from a theoretical point of view, while data availability is only considered one of the many aspects to take into account (Niemeijer 2002). Several are combinations of the two and in practice these two types are not mutually exclusive.

Based on its inventory and review, the OECD developed and pilot tested 36 water governance indicators for each of the 12 water governance principles (outlined in Fig. 2.1 above) at various scales within member countries. The full suite of indicators was released at the World Water Forum in Brasilia in March 2018. OECD is now encouraging member countries to use the various indicators; existing data and multi-stakeholder self assessment methods at various scales to apply the water governance indicators. However, water governance indicators need to be able to take into account that many water systems do not align with member country boundaries. For the most part, the frameworks and indicators included in the international review and inventory conducted by the OECD are jurisdictionally focused mostly at the national level. However, as noted in the above section, water policy and water governance are fundamentally transboundary, multi-level and subnational as many river and lake systems are not confined within political boundaries. This reality must be reflected in water governance indicators.

In the past few years there is some evidence that a new period of indicator development and application is currently developing. The recent move towards open-data by governments around the world and the global concern about the implications of climate change for water governance is facilitating a renewed interest in water governance indicators. There has been emphasis on consolidation, integration of scholarly and practitioner indicators and more multi-level and transboundary applications in complex water systems. There is also increasing recognition of the importance of context and the dynamic nature of complex water governance systems, the need to include participation and engagement indicators, emphasis on including qualitative data along with quantitative indicator data, and a focus on using key indicators within and across water systems.

The scholarship on water governance clearly indicates that water governance indicators must be able to take into account transboundary and multi-level governance factors. The scholarship on adaptive governance and adaptive capacity also clearly indicates that water governance indicators need to be able to capture the ability of water governance systems and policy regimes to adapt and including an emphasis on institutional indicators is important if water governance indicators are going to be used for diagnostic and policy analysis purposes. However, it remains to be seen if a next generation of water governance indicators is emerging. The OECD's recent efforts to develop water governance indicators related to its 12 water governance principles seems to be directed at applications within member countries, yet there is some work underway to develop and test their WGIs across complex transboundary water systems that straddle more than one member country (Johns and VanNijnatten 2018). Previous work by the OECD in its report on the "Applications of Complexity Science for Public Policy" (OECD 2009) also offers the potential to incorporate more complexity into water governance indicators work. Scholarly application of these new suite of indicators in complex water governance systems also holds a lot of promise in furthering both scholarship and practice. This however will require a more fulsome integration of the theory and concepts reviewed here and some analysis of what the scholarship brings to the table in terms of applying these indicators.

While there seems to be growing scholarly and practitioner faith in water governance indicators, at the same time, criticisms of these efforts are alive and well. Biswas and Tortajada (2010) have articulated a serious concern about water governance indicators, namely that, "it may not be possible to develop an all-purpose water governance indicator even for one country… Because governance requirements for different types of water uses are likely to be different". In addition, most applications of water governance indicators remain focused on national and subnational jurisdictions and water systems that fall within political borders. Although there have been water governance indicators developed for comparisons across countries in various regions (Hill and Engle 2013), the Asian Water Governance Index (Araral and Wang 2013) and the Mediterranean region (Burak and Margat 2016), the data collection is still primarily focused on the national level. Developing indicators for institutions, processes and networks that operate across domestic borders and applying them comparatively across complex water systems is very challenging (Saleth and Dinar 2005, Pahl-Wostl 2012). It is recognized by scholars and practitioners that dynamic, multi-scale approaches are required and that "it is more important to understand the dynamics of adaptive capacity and the relationships between common determinants in different contexts and across different scales" (Betinni et al. 2015). There is also an emerging literature that 'third-order' water governance indicators need to integrate insights from the emerging scholarship on indicator use in public policy (Bell and Morse 2013, Howlett and Cuenca 2016, IJC 2016).

More recently there have been applications of water governance indicators in comparative water systems such as river and lake systems, some of which are transboundary. A critical question for the purposes of this chapter is whether the kind of theoretical and methodological work on indicators, which have primarily been developed within countries or jurisdictions, is transferable to the transboundary context. The conceptual and methodological dynamics underlying the literature on water governance indicators reflect a growing recognition of complexity (Pahl-Wostl et al. 2012, Pahl-Wostl 2017) but, at the same time, attempts to "measure" how well we are doing with respect to water governance within countries and in transboundary contexts is challenging. Both scholars and practitioners are grappling with the application of WGIs in transboundary contexts. The next section highlights some of these applications.

Applications of Transboundary Water Governance Indicators to Date

As outlined above the literature on water governance, transboundary water governance and governance indicators is now vast and interdisciplinary. In the past decade, scholars and practitioners have been trying to move beyond general concepts of IWRM, water governance and transboundary water governance by using indicators approaches. Fundamentally, all WGIs are based on specific definitions of water governance and the methods used to collect data associated with various indicators. In this section the focus is on transboundary applications of WGIs to date. While this review is not exhaustive, what is clear from the review above is that several scholarly and practitioner models of WGIs exist, most of the applications of WGIs have been at the national scale, and those that have been used at the transboundary scale have primarily been applied to transboundary river systems.

Development of WGIs related to transboundary water systems have been around for some time. Early models of indicators used the pressure, state, impact, response indicators related to transboundary river management (Lorenz et al. 2001). Some of these indicator models included both scientific indicators and policy, governance and regulation indicators. Lorenz et al. outlined some time ago that these indicators should be applied at multiple levels (international, national, regional, and local) when being applied to international river basins. However, the authors did not apply the framework and indicators to specific transboundary river systems. Similarly, other authors have developed indicator frameworks that can be applied at various scales.

The most extensive scholarly study to date using WGIs has been conducted by Claudia Pahl-Wostl and collaborators which used 98 indicators to compare 29 within-country river systems in Europe, Latin America, Africa and Asia (Pahl-Wostl et al. 2012). This study used an analytical framework that makes a distinction between (a) water governance regime,[1] (b) regime performance and (c) environmental and socio-economic context. The focus is on empirical analysis and the links between these three elements (ibid). Under these broad categories several sets of indicators were used: formal institutional setting; regime architecture and integration; knowledge and information management; performance characteristics (progress toward policy goals, population with access, degree of public participation, transparency related to water use, surface and groundwater quality, intensity of groundwater use, water quality monitoring, etc.) and a set of context characteristics (population, state of development, water availability, water stress, land use, etc.). Data collection across all 29 cases allowed for comparative analysis of water governance regimes in river basins within both developed and developing countries in Latin America, Europe, Asia and Africa to understand how performance depends on characteristics of regimes and the context in which they are embedded. Overall, they found that regime architecture, polycentric governance regimes characterized by balancing distribution of power with effective coordination structures have higher performance and that knowledge and information management is very important, findings that were valid across diverse contexts (Pahl-Wostl et al. 2012). Although there have been some extensions of this research to 'national parts of transboundary basins' (Pahl-Wostl and Knieper 2016) the focus on WGIs in this work was on within-country water governance of complex river systems.

Van Rijswick et al. (2014) identified 10 building blocks for sustainable water governance. They argue that a three-step interdisciplinary method

---

[1] The authors indicate that "A governance regime refers to the interdependent structural features of a governance system. These include formal and informal institutions (established rules and practices) and actor networks. A governance system is a broader notion, which encompasses structural features and transient processes at both rule making and operational levels. The notion of governance takes into account the different actors and networks that help formulate and implement water policy. Governance sets the rules under which management operates. Understanding the influence of governance regime characteristics on water management is thus a prerequisite for assessing the success or failure of and providing guidance for governance reform" (Pahl-Wostl et al. 2012).

must be used to assess approaches to water shortages, water quality and floods. Combining insights from water systems analysis, economics, law and public administration they focus on assessing the strength of water management and government in terms of knowledge base, organizational processes, and implementation problems at the service level. They argue an integrated assessment methods must include a focus on 10 factors (water system knowledge; values, principles and policy discourse; stakeholder involvement; trade-offs between objectives; responsibility, authority, means; regulations and agreements; financial arrangements; engineering and monitoring; enforcement; conflict prevention and resolution). By assessing these multiple dimensions of water management and governance the authors argue that complex water systems can be assessed on a case by case basis. However, similar to Lorenz et al. they do not apply their model to transboundary cases.

Others such as de Boer et al. (2015) have advanced work on water governance by focusing on the conceptual models and the practical application of 'collaborative water governance'. They use different models to focus on multi-stakeholder participation in governance processes. They developed the 'Governance Assessment Tool' to identify the strengths and weaknesses of a specific water governance context and for designing or supporting collaborative water management in practice. They illustrate the tool's usefulness for understanding the impacts of governance on collaboration by applications in five different countries. While de Boer et al. focus on within country applications, other such as Bréthaut's (2016) focus on participatory principles and indicators that are elements of transboundary water management in the Rhone River basin highlighting again the central role of institutions and water basin authorities as central indicators related to water governance.

Applications of WGIs to Transboundary Water Systems

In terms of applications of WGIs in transboundary water systems, Milman et al. (2013) used a specific definition of WGIs to operationalize *transboundary adaptive capacity* to examine six dimensions of transboundary river basins [authority, national-level governance; common perspectives; risk planning; basin information exchange; and linkages]. For each of these they develop 12 measurable indicators related to institutional capacity and capacity to address uncertainty using existing secondary data sets such as the Political Stability Index and Government Effectiveness Index. They assessed 42 river basins that span at least two countries and have various levels of economic development, geographic features and political structures in the Middle East and Mediterranean. Using cluster analysis based on similarities between basins on the 12 indicators and an aggregate transboundary adaptive capacity index, they then develop a typology of six categories of basins: high capacity; mediated cooperation; good neighbor; dependent instability; self-sufficient and low capacity. Formal agreements, the presence of river basin organizations, trade interdependence, degree of reliance on water from the basin and data sharing were all determinants of high capacity, well-governed basins. This analysis however did not drill down into case-specific factors.

Hill and Engle (2013) and others have conducted research using *adaptive capacity indicators* in water systems at different scales. They use 'a suite of governance-related adaptive capacity indicators to investigate adaptive capacity across governance scales to explore adaptive capacity in relation to past hydrological events' (Hill and Engle 2013), including a distinction between reactive and proactive capacity. They argue that 'over the last decade there has been a growing body of literature on institutional and governance determinants and indicators of adaptive capacity in different socio-ecological systems' (Hill and Engle 2013). These scholars identify eight determinants of adaptive capacity related to water governance: economic resources, technology, information and skills, infrastructure, institutions, equity, social capital, and collective action. They elaborate that there has been wide recognition of the integrating importance of institutions and governance mechanisms for building adaptive capacity (Hill and Engle 2013) and add to their list several 'governance indicators': information and knowledge, experience and expertise, networks, transparency, trust, commitment, leadership, legitimacy, accountability, connectivity and collaboration, and flexibility. They focus on two sets of indicators: knowledge indicators and network indicators across four different water governance systems and try to determine 'enabling' and 'hindering' aspects of these indicators based on semi-structured interviews with 80 water stakeholders and experts and comparative analysis across four water governance regimes/cases: Region V in Chile, the Canton Valais in Switzerland and in two US states (Georgia and Arizona). They use WGIs broadly in a transboundary, multi-jurisdictional, regional way but not explicitly to transboundary water systems.

Garrick and DeStefano's recent work focuses on *institutional attributes of adaptive capacity* that are relevant and nested at a variety of scales from international to the community level. They develop these in the context of transboundary river systems. These indicators or 'attributes' are described as institutional design principles and enabling conditions, focusing on the necessary conditions that are required to achieve a given set of policy or performance outcomes in transboundary water systems. Drawing on Gupta and colleagues (Gupta et al. 2010) they focus on the 'inherent characteristics of institutions that enhance adaptive capacity' through the notion of an 'adaptive capacity wheel' that includes eight dimensions: (1) diversity/variety of perspectives, actors and solutions; (2) learning capacity; (3) scope for autonomous change; (4) leadership; (5) resources; (6) fair governance; (7) actor's motivation for adaptation; and (8) actor's belief in the feasibility of adaptation (Garrick and DeStefano 2016). These have been identified in their research on transboundary river basins as the critical factors associated with effective and adaptive water governance. Applications in transboundary lake systems are much more limited.

Applications of WGIs Related to Transboundary Lake Systems

It is clear from the brief review above that scholars have been trying to move beyond general concepts of water governance and transboundary water governance to develop and apply WGIs using a number of different definitions of water governance. Much of this research focuses on governance institutions and arrangements that flow from national policies or transboundary agreements. Both scholarship and practice have advanced, yet the focus remains primarily on national WGIs and the limited applications of WGIs in transboundary contexts have primarily been extended to transboundary river systems. This section reviews the use and application of WGIs to transboundary lake systems and the few studies that have tried to deploy WGIs in both transboundary lake and river systems more broadly under the category of complex transboundary water systems.

The North American Great Lakes have been an interesting laboratory for studying complex, multi-level, transboundary water governance for some time (Johns 2009, 2010, 2017, Kratnzberg and Manno 2010, de Boer and Krantzberg 2013, Grover et al. 2016). As a transboundary lake system this region faces numerous, long-standing governance challenges. It has a well-developed governance regime yet new and enduring governance challenges remain. To examine the transboundary and multi-level governance challenges in this region across water governance and a range of water governance related issues such as fisheries, invasive species nearshore governance and water use management, a team developed a conceptual and analytical framework for assessing Transboundary Governance Capacity (TGC) drawing on scholarship from international relations and law, institutional theory in political science, socio-ecological systems theory, and governance theory in public policy and public administration (VanNijnatten et al. 2016).

The authors focus on five indicators or attributes that they argue must be assessed in transboundary lake systems: leadership, necessary and sufficient participation of key stakeholders and users, shared discourse and mutual understanding, sustainable resources and a strong institutional basis (VanNijnatten et al. 2016). The authors place particular emphasis on institutional WGIs based on the argument that the presence of a strong institutional basis, while not necessarily *more* important than the other attributes, is *foundational* in terms of TGC, as institutions act to channel policy discourse, structure policy choices and resources, and provide opportunities or constraints for policy actors in terms of achieving outcomes.

Using this analytical framework, the authors demonstrated the value of focusing on institutional WGIs, while also the challenge of comprehensively looking at the set of five key WGIs they identified from the literature. The application of their model across various water-related environmental issues however demonstrated the value of comparing WGIs across policy domains. In addition, applications of the institutional WGIs to other transboundary water systems in the Arctic (Freidman 2016), the Columbia River basin in the US and the Murray Darling basin in Australia (Garrick et al 2016) illustrated the value of focusing on institutional WGIs as a starting point for more comprehensive WGI research. Other complementary studies have also illustrated the significance of shared discourse and understanding. A study by Song et al. (2015) indicated that scientific capacity and collaboration is also an important indicator to consider when defining and measuring water governance in this region (Song et al. 2015).

In addition to scholarly studies applying WGIs to transboundary lake systems, and one not captured in the OECD's inventory, is the *Transboundary Water Assessment Program* (TWAP). The TWAP is a practitioner initiative that has its foundations in work from the Global Environment Facility (GEF) a coalition originally established in the 1990s by the World Bank in partnership with the UNDP and UNEP related to several UN Conventions. In 2009–10 this cluster of organizations started the TWAP. The first step was to take stock of transboundary water systems across the globe. Part of this project focused on transboundary lakes using spatial analysis of primarily NASA and USGS global-scale databases to generate a list of 1600 transboundary lakes and reservoirs. Using several criteria, the list was then reduced to 204 transboundary lakes and reservoirs, including 33 in Africa, 51 in the Asia region, 30 in South America, 70 in the European region, and 20 in North America. A final list of 53 priority transboundary lakes was selected for more detailed scenario and indicator analysis (23 lakes in Africa, 8 in Asia, 9 in Europe, 6 in South America, and 7 in North America). The TWAP consists of five independent indicator-based assessments and the linkages between them, including their socioeconomic and governance-related features. The

2016 Report focuses on two sets of indicators related to transboundary lake systems: water security threats and biodiversity threats. Using these indicators, the TWAP ranked the 53 priority transboundary systems in terms of threats (TWAP 2016). The most highly ranked lakes in terms of threats were in Africa, the least threatened transboundary lakes were in North America. This review illustrates how important the definition of WGIs is in the analysis of transboundary lake systems.

In summary, this evaluation of WGI applications in transboundary water systems reveals that despite different disciplinary foundations definitions of water governance and transboundary water governance are the key to WGIs. The focus of WGIs can vary related to water security, water threats, water stress, water quantity, or water quality. The focus also varies depending on whether water governance and TWG are defined as systems that are being assessed in past or present context or whether the WGIs are defined in a more dynamic way in terms of a TWG system being able to adapt to change or uncertainty. A review of applications of WGIs at the transboundary scale also highlights that despite these definitional differences there are some common foundations to WGIs. Most applications include some focus on the institutional and policy foundations of WGIs; some measures of engagement of stakeholders and water users; and some WGIs related to knowledge at the system or actor levels. Finally, the chapter highlights that most applications of WGIs have been within countries and those that have focused on transboundary water systems have primarily focused on transboundary river systems, with only a few applications of WGIs to complex, transboundary lake systems involving multiple countries and jurisdictions.

Conclusions

This chapter outlines how both scholars and practitioners have developed and used WGIs to analyze and assess water governance and policy in a wide range of contexts and scales. Research on transboundary lake systems reveals that the focus can vary from WGIs based on water security and scarcity to those focused on water quality. Applications can also have both a transboundary and domestic focus. The chapter also outlines how important disciplinary and intellectual foundations are for WGIs and that definitions of both water governance and transboundary water governance matter in terms of WGIs. Applications of WGIs to date clearly indicate institutional factors, resources, engagement, and knowledge are critical for assessing and improving water governance and policy outcomes.

Despite some progress in applying WGIs within countries, it is evident that studies using WGIs in transboundary cases have primarily focused on transboundary river systems. Studies using WGIs related to transboundary lake systems are less common. Nonetheless, when combined with studies using WGIs within countries it is evident that there are parallels in terms of the types of WGIs used. Studies in transboundary systems have added a multi-level governance focus and revealed that much can be learned from transboundary cases and comparison of water governance across cases. Transboundary lake cases indicate sets of WGIs should include a focus on institutions, the role of formal and informal networks, resources, stakeholder engagement, and the knowledge systems that support the governance system. The development and application of these sets of WGIs can be valuable in terms of identifying and addressing implementation and engagement deficits (McLaughlin and Krantzberg 2011, 2012, Krantzberg et al. 2015). There is well-developed literature on the important role of institutions in water governance at all scales indicating that WGIs should include a focus on formal state institutions but also formal and informal networks of organizations and actors involved in policy implementation.

Indicators related to resources are also important including financial, personnel, information and knowledge resources. Collecting data on resource allocations and personnel associated with various water governance systems remains challenging, particularly in transboundary water systems. However, some water governance research has found that knowledge attributes of policy systems and actors, particularly those with a future orientation are important (Hill and Engle 2013, Hill 2015). Comparative analysis by Milman et al. 2013 across 42 transboundary river basins using macro-level indicators and existing datasets found there is a "paucity of mechanisms related to knowledge and addressing uncertainty" and that those transboundary water systems with data sharing and shared water norms had high levels of transboundary adaptive capacity related to climate change. This work indicates that WGIs should incorporate some measure of ability to generate and use scientific knowledge, and some measure of the ability of water governance 'systems' and institutions to generate and use knowledge to adapt to uncertainty.

Scholarship on water governance also points to the engagement of water users and stakeholders as critically important to both policy effectiveness and democratizing water governance systems. There is clear evidence that governance and policy systems that engage key water users and stakeholders are more successful, particularly at local and watershed scales (Ostrom 2007). Engagement indicators can also be valuable for identifying water users and stakeholders who have not been engaged in the formal water governance regimes. As the Great Lakes case illustrates, there is clearly an opportunity to more fully engage Indigenous knowledge and communities (Norman 2015, Norman and Bakker 2016). Several scholars have called for more systematic analysis of engagement indicators and cooperative structures across water governance systems (Pahl-Wostl et al. 2012) using WGIs. Scholarship by deBoer et al. (2013, 2016) using within-country cases demonstrates that engagement factors including trust, collective responsibilities and shared resources are important. Edelenbos and van Meerkerk's (2015) research on 'connective capacity' also demonstrates the significance of boundary spanning leadership and trust between actors. Research using stakeholder and Social Network Analysis (SNA) is emerging as a critical frontier of water governance and engagement research (Lienert et al. 2013, Johns 2015, Ingold et al. 2016) but network indicators have yet to become part of WGIs. The OECD is also focused on stakeholder engagement as one of its 12 water governance principles and in its work on indicators (OECD 2015c, OECD 2018). It is clear that WGIs should thus include some indicators related to stakeholder engagement and the engagement of the multiple users of water in transboundary lakes and other transboundary systems.

It remains to be seen if the OECD WGIs can be used more broadly, applied to transboundary lake systems, and if they can be used as the basis of more scholarly and practitioner collaboration in the future. A current project is underway in Canada to see if the OECD's WGIs and a suite of knowledge and engagement indicators can be applied as part of a comparative analysis of the North American Great Lakes region as a transboundary lake system and the Rio Grande region as a transboundary river system (Johns and VanNijnatten 2018).

Despite the significant progress on water governance indicators in the past decade, this chapter highlights that significant challenges remain related to developing and applying water governance indicators in transboundary lake systems. First, there are conceptual challenges as the disciplinary perspective and way scholars and practitioners define water governance and transboundary water governance has implications for the development and operationalization of WGIs. Second, enduring challenges of measurement and data collection related to indicators remains. There are long-standing concerns about the use of indicators to simplify complex social, political, economic and environmental systems. This is even more challenging given the critique by some scholars that a focus on water governance regime factors is not enough and that there needs to be a focus on broader forces and pressures, such as adaptation to climate change. Third, there are challenges related to the ultimate purpose and use of WGIs. Most scholarly and practitioner work on WGIs is based on some faith that knowledge about water governance systems will help us use, govern, and manage water more justly and sustainably. There is a fundamental assumption underpinning WGIs that only through better governance systems can we address the current water challenges in domestic and international contexts. Scholarship in public policy and public administration clearly indicates the use of indicators related to policy making, implementation, evaluation, policy change and governance reforms is a significant challenge. WGIs need to integrate insights from the emerging scholarship on indicator use in public policy (Howlett and Cuenca 2016). This may be even more challenging at the transboundary scale if users of WGIs are not easily identifiable and part of the indicator development process.

In summary, this chapter outlines that there are both opportunities and challenges related to using WGIs to improve our understanding and improve governance and policy outcomes in transboundary lake systems. If scholars and practitioners can focus more of their efforts on transboundary lake systems and overcome some of the challenges that exist related to WGIs, the prospects for developing and using WGIs to improve transboundary lake governance are promising. Advancing the use of WGIs to enhance governance of transboundary lakes will require interdisciplinary collaborations between scholars and practitioners to prioritize the use of existing WGIs or the development of WGIs for transboundary lake systems and involve: clearly outlining what the purpose of developing and using WGIs is; clearly defining the central concepts of good governance, water governance and transboundary water governance; outlining the assumptions that underpin WGIs; recognizing the inherent tension between indicators and embracing complexity; valuing both quantitative and qualitative measures and data collection methods, and recognizing that WGIs are only one set of tools that can be used to improve transboundary lake governance in the future.

References

Armitage, D., Béné, C., Charles, A.T., Johnson, D. and Allison, E.H. (2012). The interplay of well-being and resilience in applying a social-ecological perspective. *Ecology and Society*, 17(4):

Araral and Wang. (2013). Water Governance 2.0: A review and second generation research agenda. *Water Resources Management*, 27(11): 3945–3957.

Azar, C., J. Holmberg and C. Lindgren. (1996). Socio-ecological indicators for sustainability. *Ecological Economics*, 18: 89–112.

Bell, S. and Morse, S. (2013). Toward an understanding of how policy making groups use indicators. *Ecological Indicators*, 35: 13–23.

Betinni, Y., Brown, R. and deHann, F. (2015). Exploring institutional adaptive capacity in practice: examining water governance adaptation in Australia. *Ecology and Society*, 20(1): https://www.ecologyandsociety.org/vol20/iss1/art47/.

Biswas, A.K. and Tortajada, C. (2010). Future water governance: problems and perspectives. *International Journal of Water Resources Development*, 26(2): 129–139.

Biswas, A. (2008). Integrated water resource management: is it working? *International Journal of Water Resources Development*, 24(1): 5–22.

Blomquist, W. and Elinor Ostrom. (1985). Institutional capacity and the resolutions of a commons dilemma. *Review of Policy Research*, 5(2): 383–393.

Breitmeier, H., Young, O. and Zurn, M. (2006). Analyzing International Environmental Regimes. Cambridge: MIT Press.

Bressers, Hans and Kris Lulofs (eds.). (2010). Governance and Complexity in Water Management: Creating Cooperation through Boundary Spanning Strategies (Cheltenham, UK: Edward Elgar).

Bréthaut Christian. (2016). River management and stakeholders' participation: the case of the Rhone River, a fragmented institutional setting. *Environmental Policy and Governance*, 26(4): 292–305.

Burak, Selmin and Jean Margat. (2016). Water management in the mediterranean region: concepts and policies. *Water Resource Management*, 30(15): 5779–5797.

Cash, D.W., Adger, W., Berkes, F., Garden, P., Lebel, L., Olsson, P., Pritchard, L. and Young, O. (2006). Scale and cross-scale dynamics: governance and information in a multilevel world. *Ecology and Society*, 11(2): 8.

Chaffin, B., Gosnell, H. and Cosens, B. (2014). A decade of adaptive governance scholarship. *Ecology and Society*, 19(3): https://www.ecologyandsociety.org/vol19/iss3/art56/.

Cox, M., Villamayor-Tomás, S., Epstein, G., Evans, L., Ban, N., Fleischman, F., Nenadovic, N. and Garcia-Lopez, G. (2016). Synthesizing theories of natural resource management and governance. *Global Environmental Change*, 39: 45–56.

Dale, V.H. and Beyeler, S.C. (2001). Challenges in the development and use of ecological indicators. *Ecological Indicators*, 1: 3–10.

Dunn, G. and Bakker, K. (2009). Canadian approaches to assessing water security: an inventory of indicators. Policy Report: Fostering Water Security in Canada Project. University of British Columbia: http://www.watergovernance.ca/PDF/IndicatorsReportFINAL2009.pdf.

deBoers, Cheryl and Gail Krantzberg. (2013). Great lakes water governance: A transboundary inter-regime analysis. pp. 315–332. *In*: Jurian Edelenbos, Nanny Bressers and Peter Scholten (eds.). Water Governance as Connective Capacity (New York: Routledge).

de Boer, Cheryl, Joanne Vinke-de Kruijf, Gül Ozerol and Hans Bressers (eds.). (2013). Water Governance, Policy and Knowledge Transfer: International Studies on Contextual Water Management. New York: Routledge.

de Boer, Cheryl, Joanne Vinke-de Kruijf, Gul Ozerol and Hans Bressers. (2016). Collaborative Water Resource Management: What makes up a supportive governance system? *Environmental Policy and Governance*, 26(4): 229–241.

de Loë, R.C. and Patterson, J.J. In Press. Rethinking water governance: moving beyond water-centric perspectives in a connected and changing world. *Natural Resources Journal*, 57(1).

De Stefano, Lucia. (2010). International initiatives for water policy assessment: a review. *Water Resources Management*, 24: 2449–2466.

Earle, Anton, Jagerskog, Andres and Ojendal, Joakim (eds.). (2010). Transboundary Water Management: Principles and Practice (New York: Earthscan).

Edelenbos, J. and van Meerkerk, I. (2015). Connective capacity in water governance practices: the meaning of trust and boundary spanning for integrated performance. *Current Opinion in Environmental Sustainability*, 12: 25–29.

Edelenbos, J. and Teisman, G. (2013). Water governance capacity: the art of dealing with a multiplicity of levels, sectors, and domains. *International Journal of Water Governance*, 1: 89–108.

Edelenbos, Jurian, Nanny Bressers, and Peter Scholten. (2013b). Water Governance as Connective Capacity. New York: Routledge.

Engle, N.L. (2011). Adaptive capacity and its assessment. *Global Environmental Change*, 21(2): 647–656.

Farrell, A. and Hart, M. (1998). What does sustainability really mean? The search for useful indicators. *Environment: Science and Policy for Sustainable Development*, 40(9): 4–31.

Folke, C., Hahn, T., Olsson, P. and Norberg, J. (2005). Adaptive governance of social-ecological systems. *Annual Review of Environment and Resources*, 441–473.

Friedman, Kathryn Bryk. (2016). Institutions and Transboundary Governance Capacity (TGC) in the Arctic: Insights from the TGC. *Framework International Journal of Water Governance*, 4(8): 133–154.

Garrick, Dustin and Hall, J. (2014). Water security and society: risks, metrics and pathways. *Annual Review of Environment and Resources*, 39: 611–39.

Garrick, D. (2015). Water Allocation in Rivers Under Pressure: Water Trading, Transaction Costs and Transboundary Governance in the Western US and Australia. Cheltenham, UK: Edward Elgar Publishing.

Garrick, Dustin, Gail Krantzberg and Savitri Jetoo. (2016). Building transboundary water governance capacity for non-point pollution: A comparison of Australia and North America. *International Journal of Water Governance*, 4(8): 111–132.

Garrick, Dustin and De Stefano, Lucia. (2016). Adaptive capacity in federal rivers: coordination challenges and institutional responses. *Current Opinion in Environmental Sustainability*, 21: 78–85.

Garriga, R.G. and Foguet, A. (2015). The Water Poverty Index: Assessing Water Scarcity at Different Scales, Paper Presented at Congres UPC Sostenible.

George, Frederickson, Kevin, B. Smith, Christopher, W. Larimer and Michael, J. Licari (eds.). (2016). Theories of governance. In The Public Administration Theory Primer (Boulder CO: Westview Press), 222–248.

Gunderson, L. and Light, S. (2006). Adaptive management and adaptive governance in the everglades ecosystem. *Policy Sciences*, 39: 323–334.

Gupta, J., Termeer, C., Klostermann, J., Meijerink, S., van den Brink, M., Jong, P., Nooteboom, S. and Bergsma, E. (2010). The adaptive capacity wheel: a method to assess the inherent characteristics of institutions to enable the adaptive capacity of society. *Environmental Science & Policy*, 13: 459–471.

Grover, Velma and Gail Krantzberg. (2014). Transboundary water management: Lessons learnt from North America. *Water International*, online.

Hill, Clarvis, Margot, Erin Bohensky and Masaru Yarime. (2015). Can resilience thinking inform resilience investments? Learning from resilience principles for disaster risk reduction. *Sustainability*, 7(7): 9048–66.

Hill, Margot and Nathan, L. Engle. (2013). Adaptive capacity: tensions across scales. *Environmental Policy and Governance*, 23(3): 177–192.

Ho, Selina. (2017). Introduction to transboundary river cooperation: actors, strategies and impact. *Water International*, 42(2): 97–104.

Holling, C.S. (2001). Understanding the complexity of economic, ecological and social systems. *Ecosystems*, 4: 390–405.

Howlett, M. and Cuenca, J. (2016). The use of indicators in environmental policy appraisal: Lessons from the design and evolution of water security policy measures. *Journal of Environmental Policy and Planning*, online.

Huitema, D., Mostert, E., Egas, W., Moellenkamp, S., Pahl-Wostl, C. and Yalcin, R. (2009). Adaptive water governance: assessing the institutional prescriptions of adaptive (co) management from a governance perspective and defining a research agenda. *Ecology and Society*, 14(1): 26.

Huntjens, P., Leel, L., Pahl-Wostl, C., Camkin, J., Schulze, R. and Kranz, N. (2012). Institutional design propositions for the governance of adaptation to climate change in the water sector. *Global Environmental Change*, 22(1): 67–81.

Ingold Karen, M. Fischer, C. de Boer and P. Mollinga. (2016). Water management across borders, scales and sectors: Recent developments and future challenges in water policy analysis. *Environmental Policy and Governance*, 26(4): 223–228.

International Joint Commission. (2011). Assessment of Progress Made Towards Restoring and Maintaining Great Lakes Water Quality Since 1987, IJC Ottawa/Washington.

International Joint Commission. (2016). An Assessment of the Communicability of the International Joint Commissions Ecosystem Indicators and Metrics, Great Lakes Science Advisory Board, Communication Indicator Workgroup.

Jager, N. (2015). Transboundary cooperation in European water governance—a set-theoretic analysis of international river basins. *Environmental Policy and Governance*, 26(4): 278–91.

Jägerskog, A. and Zeitoun, M. (2009). Getting Transboundary Water Right: Theory and Practice for Effective Cooperation. Report Nr. 25. SIWI, Stockholm.

Johns, Carolyn. (2017). The great lakes, water quality and water policy in Canada. *In*: Steven, Renzetti and Diane, Dupont (eds.). Water Policy and Governance in Canada, Springer Publications, 159–178.

Johns, Carolyn. (2015). Opportunities and Challenges of Using Social Network Analysis in Public Policy Research: Analyzing Governance and Policy Implementation Challenges in the Great Lakes Region. Paper presented at International Conference on Public Policy, Milan, Italy, July 2, 2015.

Johns, C. (2009). Water pollution in the Great Lakes Basin: the global-local dynamic. pp. 95–129. *In*: Christopher Gore and Peter Stoett (eds.). Environmental Challenges and Opportunities: Local-Global Perspectives on Canadian Issues (Toronto: Emond Montgomery).

Johns, C. (2010). Transboundary water pollution efforts in the Great Lakes: the significance of national and sub-national policy capacity. pp. 63–82. *In*: Rabe, B. and S. Brooks (eds.). *Environmental Governance on the 49th Parallel: New Century, New Approaches*, Washington, DC: Woodrow Wilson International Center for Scholars, Canada Institute.

Johns, Carolyn and Debora VanNijnatten. (2018). Embracing Complexity and Adaptability: Comparative Analysis and Key Indicators for Improving Transboundary Water Governance, project funded by the Social Sciences and Humanities Research Council of Canada, http://greatlakespolicyresearch.org/about.

Kaufmann, Daniel, Art Kraay and Massimo Mastruzzi. (2005). Governance Maters IV: Governance Indicators 1996–2004, Policy Research Working Paper Series, 3630, The World Bank.

Kaufmann, Daniel and Aart Kraay. (2008). Governance indicators: where are we, where should we be going? *World Bank Research Observer*, 23: 1–30.

Kaufmann, D., Kraay, A. and Mastruzzi, M. (2009). Governance Matters VIII: Aggregate and Individual Governance Indicators 1996–2008. Policy Research Working Paper 4978, World Bank, Development Group, Macroeconomics and Growth Team. Available at: http://www-wds.worldbank.org/external/default/WDSContentServer/WDSP/IB/2009/06/29/000158349_20090629095443/Rendered/PDF/WPS4978.pdf.

Knieper, Christian and Pahl-Wostl, Claudia. (2016). HYPERLINK "https://EconPapers.repec.org/RePEc:spr:waterr:v:30:y:2016:i:7:d:10.1007_s11269-016-1276-z" A comparative analysis of water governance, water management, and environmental performance in river basins. *Water Resources Management: An International Journal*, 30(7): 2161–2177.

Krantzberg, Gail and Jack, P. Manno. (2010). Renovation and innovation; it's time for the Great Lakes regime to respond. *Water Resources Management*, 24(15): 4273–4285.

Krantzberg, Gail, Irena, F. Creed, Kathryn, B. Friedman, Katrina, L. Laurent, John, A. Jackson, Joel Brammeier and Donald Scavia. (2015). Community engagement is critical to achieve a "thriving and prosperous" future for the Great Lakes–St. Lawrence River basin. *Journal of Great Lakes Research*, 41 Supplement 1: 188–191.

Langhans, S.D., Reichert, P. and Schuwirth, N. (2014). The method matters: a guide for indicators aggregation in ecological assessments. *Ecological Indicators*, 45: 494–507.

Lienert, Judit, Florian Schnetzer and Karin Ingold. (2013). Stakeholder Analysis combined with social network analysis provides fine-grained insights into water infrastructure planning process. *Journal of Environmental Management*, 125: 134–148.

Lorenz, Carolin M., Alison J. Gilbert and W.P. Cofino. (2001). Indicators for transboundary river management. *Environmental Management*, 28(1): 115–129.

Luh, Jeanne, Rachel Baum and Jamie Bartram. (2013). Equity Index in water and sanitation. *International Journal of Hygiene and Environmental Health*, 216(6): 662–671.

Medema, W., McIntosh, B. and Jeffrey, P. (2008). From premise to practice: a critical assessment of integrated water resource management and adaptive management approaches in the water sector. *Ecology and Society*, 13(2): 29.

McLaughlin, Chris and Gail Krantzberg. (2011). An appraisal of policy implementation deficits in the great lakes. *Journal of Great Lakes Research*, 37: 390–396.

McLaughlin, Chris and Gail Krantzberg (2012). An appraisal of management pathologies in the great lakes. *Science of the Total Environment*, 40–47.

Milman, Anita, Lisa Bunclark, Declan Conway and William Neil Adger. (2013). Assessment of institutional capacity to adapt to climate transboundary river basins. *Climate Change*, 121: 775–770.

Molle, F. and Mollinga, P. (2003). Water poverty indicators: conceptual problems and policy issues. *Water Policy*, 5: 529–544

Morris, M. and de Loë, R.C. (2016). Cooperative and adaptive transboundary water governance in Canada's Mackenzie River Basin: status and prospects. *Ecology and Society*, 21(1): 26.

Moss, T. and Newig, J. (2010). Multilevel water governance and problems of scale: setting the stage for a broader debate. *Environmental Management*, 46: 1.

Muriithi, K.J. Margarita, Jannin, N., Sajid, N., Sahibjeet, S. and Sudhanshu, S. (2015). Quantifying Governance: An Indicators Approach, London School of Economics.

Niemeijer, D. (2002). Developing indicators for environmental policy: data-driven and theory-driven approaches examined by example. *Environmental Science and Policy*, 4(2): 91–103.

Norman, E.S., Dunn, G., Bakker, K., Allen, D.A. and Cavalcanti de Albuquerque, R. (2013). Water security assessment: integrating governance freshwater indicators. *Water Resources Management*, 27(2): 535–51.

Norman, Emma. (2015). Governing Transboundary Waters: Canada, the United States and Indigenous Communities. New York: Routledge.

Norman, E., Dunn, G., Bakker, K., Allen, D.M. and Cavalcanti de Albuquerque, R. (2013). Water security assessment: integrating governance and freshwater indicators. *Water Resources Management*, 27(2): 535–51.

Norman, E. and K. Bakker. (2016). Transcending borders through postcolonial water governance? Indigenous water governance across the Canada-US Border. *In*: Renzetti, S. and Dupont, D. (eds.). Water Policy and Governance in Canada (New York: Springer).

OECD. (2018). Implementing the OECD Principles on Water Governance Indicator Framework and Evolving Practices, OECD Studies on Water, Paris.

Organization for Economic Cooperation and Development (OECD). (2008). OECD Key Environmental Indicators, OECD Environment Directorate, Paris, France.

Organization for Economic Cooperation and Development (OECD), Global Science Forum. (2009). Applications of Complexity Science for Public Policy.

Organization for Economic Cooperation and Development (OECD). (2011). Water Governance in OECD Countries: A Multilevel Approach.

Organization for Economic Cooperation and Development (OECD). (2015a). Inventory of Existing Frameworks for Assessing Water Governance. Water Governance Initiative, Paris.

Organization for Economic Cooperation and Development (OECD). (2015b). Water Governance Principles. Water Governance Initiative, Paris.

Organization for Economic Cooperation and Development (OECD). (2015c). Stakeholder Engagement in Water Governance: Towards Indicators. Water Governance Initiative.

Ostrom, E. (1990). Governing the Commons: The Evolution of Institutions for Collective Action. Cambridge University Press.

Ostrom, E. (2007). The governance challenge: matching institutions to the structure of social ecological systems. *In*: S. Levin (ed.). The Princeton Guide to Ecology. Princeton, NJ; Princeton UP.

Ostrom, E. and Cox, M. (2010). Moving beyond panaceas: a multi-tiered diagnostic approach for socio-ecological analysis. *Environmental Conservation*, 37(4): 451–63.

Pahl-Wostl. (2007). Transitioning toward adaptive management of water facing climate and global change. *Water Resources Management*, 21: 49–62.

Pahl-Wostl, C. (2009). A conceptual framework for analyzing adaptive capacity and multi-level learning processes in resource governance regimes. *Global Environmental Change*, 19: 354–365.

Pahl-Wostl, C., Holtz, G., Kastens, B. and Kneiper, C. (2010). Anayzing complex water governance regimes: the management and transition framework. *Environmental Science and Policy*, 13: 571–581.

Pahl-Wostl, C., Lebel, L., Knieper, C. and Nikitina, E. (2012). From applying panaceas to mastering complexity: toward adaptive water governance in river basins. *Environmental Science and Policy*, 23: 24–34.

Pahl-Wostl, Claudia. (2017). An evolutionary perspective on water governance: from understanding to transformation. *Water Resources Management*, 31: 2917–2932.

Plummer, R., Armitage, D.R. and deLoe, R. (2013). Adaptive co-management and its relationship to environmental governance. *Ecology and Society*, 29(2): 209–227.

Rogers, P. and Hall, A.W. (2003). Effective Water Governance. TEC Background Papers No. 7, Global Water Partnership, Stockholm. https://www.gwp.org/globalassets/global/toolbox/publications/background-papers/07-effective-water-governance-2003-english.pdf.

van Rijswick, Marleen, Jurian Edelenbos, Petra Hellegers, Matthijs Kok and Stefan Kuks. (2014). Ten building blocks for sustainable water governance: an integrated method to assess the governance of water. *Water International*, 39(5): 725–742.

Saleth, R.M. and Dinar, A. (2005). Water institutional reforms: theory and practice. *Water Policy*, 7(1): 1–19.

Song, Andrew, M., Gordon Hickey, Owen Temby and Gail Krantzberg. (2015). Assessing transboundary scientific collaboration in the Great Lakes of North America. *Journal of Great Lakes Research*, online article.

Stokke, O. (2012). Disaggregating International Regimes: A New Approach to Evaluation and Comparison. Cambridge: MIT Press.

Suhardiman, Diana and Mark Giordano. (2012). Process-focused analysis in transboundary water governance research. *International Environmental Agreements*, 12(3): 299–308.

Sullivan, Caroline. (2002). Calculating a water poverty index. *World Development*, 30(7): 1195–1210.

Transboundary Water Assessment Program (TWAP) Global Environment Facility. (2016). Transboundary Water Assessment Project. http://www.geftwap.org/twap-project.

Tunstall, D.B. (1979). Developing indicators of environmental quality: the experience of the council on environmental quality, *Social Indicators Research*, 6: 301–347.

United Nations, UN Water. (2008). International Decade for Action: Water for Life 2005–2015. http://www.un.org/waterforlifedecade/water_cooperation.shtml.

United Nations, UN Water (2016). Transboundary Waters. http://www.un.org/waterforlifedecade/transboundary_waters.shtml.

United Nations Environment Program (UNEP) and International Lake Environment Committee. (2016). Transboundary River Basins: Status and Trends, Summary for Policy Makers. United Nations Environment Programme (UNEP), Nairobi.

United Nations, UN Water. (2018). Water Action Decade 2018–2028. http://www.wateractiondecade.org.

VanNijnatten, Debora, Carolyn Johns, Kathryn Friedman and Gail Krantzberg. (2016). Assessing adaptive transboundary governance in the great lakes basin: the role of institutions and networks. *International Journal of Water Governance*, 16(1): 7–32.

Van Rijswick, Marleen, Jurian Edelenbos and Petra Hellegers. (2014). Ten building blocks for sustainable water governance: an integrated method to assess the governance of water. *Water International*, 39(5): 725–742.

Verwayen, H. (1984). Social indicators: actual and potential uses. *Social Indicators Research*, 14(1): 1–27.
Vincent, K. (2007). Uncertainty in adaptive capacity and the importance of scale. *Global Environmental Change*, 17(1): 12–24.
Wiek, Arnim and Kelli L. Larson. (2012). Water, people, and sustainability—a systems framework for analyzing and assessing water governance regimes. *Water Resource Management*, 26: 3153–3171.
Yale Environmental Performance Index (2016). *Global Metrics for the Environment*. https://issuu.com/2016yaleepi/docs/epi2016_final.
Young, O. (2001). Inferences and indices: Evaluating the effectiveness of international environmental regimes. *Global Environmental Politics*, 1(1): 99–121.
Zeitoun, M., Mirumachi, N. and Warner, J. (2011). transboundary water interaction: the influence of soft power. *International Environmental Agreements*, 11: 159–178.

CHAPTER 3

Groundwater Governance and Assessment in a Transboundary Setting

Sharon B. Megdal[1,*] *and Jacob D. Petersen-Perlman*[2]

Introduction

Reliance on groundwater is increasing across the globe. Groundwater constitutes more than 95% of the Earth's unfrozen freshwater and is integral part of the water cycle (Alley et al. 2016a). With the invention of the centrifugal pump, advances in deep well pumping technology, increases in global population, and an increase in the standard of living throughout the world, groundwater has become more relied-upon as a water source for many communities, irrigators, and industries (Jarvis 2010). However, this increased reliance upon groundwater as a source for water has become problematic due to a series of issues associated with its overuse. Because of groundwater being physically invisible, it can be difficult to quantify reductions of water in storage or increases in depth to water. Likewise, contamination is hard to detect and often even harder to remedy.

There is a growing body of scholarship examining groundwater governance approaches throughout the world (McCaffrey 1998, Puri 2001, Eckstein and Eckstein 2005, Puri and Aureli 2005, Jarvis 2008, Braune and Adams 2013, Megdal et al. 2015, Sugg et al. 2015, Varady et al. 2016).

---

[1] Director, Water Resources Research Center. University of Arizona. 350 N. Campbell Avenue, Tucson, Arizona, USA 85719.
[2] Research Analyst, Water Resources Research Center. University of Arizona. 350 N. Campbell Avenue, Tucson, Arizona, USA 85719.
 Email: jacobpp@email.arizona.edu
* Corresponding author: smegdal@email.arizona.edu

Scholars have proposed new international legal principles that account for groundwater (McCaffrey 1998) and strategies for groundwater conflict resolution and management (Jarvis 2008). In general, scholars have observed that institutional structures for resolving problems associated with groundwater are underdeveloped, resolution processes are lengthy and piecemeal, and that even the few successful examples of groundwater governance involve institutional improvisation rather than systematic application of theories of desirable institutional structures (Blomquist and Ingram 2003, Knüppe 2011).

When interjurisdictional boundaries are introduced, the complications of groundwater governance increase. Transboundary groundwater basins have historically received significantly less attention compared to transboundary river basins (Puri 2001, Eckstein and Eckstein 2005, Puri and Aureli 2005). Moreover, while it is well known that aquifers cross international boundaries and that groundwater often contributes to transboundary river systems, transboundary surface and groundwater systems are often governed and managed in separate regimes. Complicating this further, the hydrologic relationships between surface water systems and groundwater systems are often not very well-understood (Condon and Maxwell 2015), making the task of determining the basic sovereignty of groundwater resources hard to solve, unaddressed, or unasked (Jarvis et al. 2005).

There are almost 600 known transboundary aquifers that have been mapped since 2003 by the International Groundwater Resources Assessment Centre (IGRAC 2015, Fig. 3.1), an initiative of UNESCO and

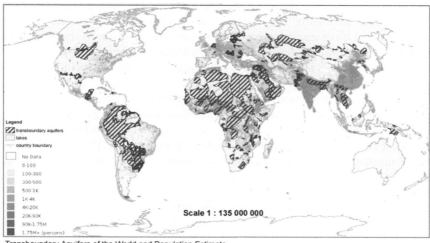

Transboundary Aquifers of the World and Population Estimate
(Source: Socioeconomic Data and Applications Center; layer name: Population Count Future Estimate 2015)

Figure 3.1. Transboundary aquifers of the world and population. Source: IGRAC 2015.

the World Meteorological Association (Sanchez et al. 2016). Recent studies (Taylor et al. 2013, Döll et al. 2014, Famiglietti 2014) suggest that humans are increasingly relying upon nonrenewable groundwater resources across the world, leading to further pressures on transboundary aquifers.

In this chapter, the authors first focus on international and regional efforts made to increase groundwater's visibility. The authors then provide an overview of groundwater governance in practice. This is then followed with a discussion of groundwater governance in a transboundary setting, including consideration of commonly accepted principles for governing groundwater and recent developments in legal principles for transboundary groundwater governance. The authors then consider a case study of groundwater assessment along the United States (U.S.)-Mexico border in the context of these principles and explain the value of the cooperative guiding framework established for this assessment.

Increasing Groundwater's Visibility and Groundwater Governance Principles

Scholars have observed that institutions charged with governing groundwater have undeveloped or fragmented structures for resolving critical problems (Giordano and Villholth 2007, Giordano 2009, Theesfeld 2010, de Chaisemartin et al. 2017). Because of these problems, effective groundwater governance is essential. For the discussion, the authors define groundwater governance as "the overarching framework of groundwater use laws, regulations, and customs, as well as the processes of engaging the public sector, the private sector, and civil society" (Megdal et al. 2015).

Because problems associated with groundwater governance and management have historically received limited attention in the literature, intergovernmental and academic organizations have recently focused on increasing the visibility of groundwater with the aim to improve its governance and management. Some of the more notable efforts involving academicians, government officials, and other expert practitioners are as follows: (1) the international project "Groundwater Governance: A Global Framework for Action" funded by The Global Environmental Facility (GEF); (2) the "Groundwater Visibility Initiative" (GVI) workshop of the American Water Resources Association (AWRA) and the National Groundwater Association (NGWA) and associated published articles; and (3) the International Symposium on Managed Aquifer Recharge (ISMAR9) "Call to Action: Sustainable Groundwater Management Policy Directives" (ISMAR9 Policy Declaration). These three efforts each produced principles for groundwater management, planning, and assessment, which the authors have categorized and summarized in Table 3.1.

Groundwater Governance and Assessment in a Transboundary Setting 43

Table 3.1. Principles for groundwater planning, management, and assessment created by three efforts to increase groundwater visibility.

Effort / Principle	Groundwater Governance: A Global Framework for Action	Groundwater Visibility Initiative	ISMAR9 Policy Declaration
Stakeholder engagement and inclusion	There are appropriate and implemented legal, regulatory and institutional frameworks for groundwater that establish public guardianship and collective responsibility, permanent engagement of stakeholders and beneficial integration with other sectors, including other uses of the subsurface space and its resources.	Governing and managing groundwater requires working with people.	Effective groundwater management requires collaboration, robust stakeholder participation and community engagement.
Proper assessment and data for analysis	All major aquifers are properly assessed, and the resulting information and knowledge are available and shared, making use of up-to-date information and communication techniques.	Data and information are key for increasing knowledge for water withdrawals and consumptive use. Some "secrets" remain—science needs to improve understanding of climate impacts on both supply and demand for groundwater and its interaction with surface water.	Aquifer systems are unique, need to be well understood, and groundwater should be invisible no more.
Management and planning	Groundwater management plans are prepared and implemented for the priority aquifers. Groundwater management agencies, locally, nationally and internationally, are resourced and their key tasks of capacity building, resource and quality monitoring, and promoting demand management and supply-side measures are secured.	We need to take care of what we have. To be robust, agriculture, energy, environment, land-use planning, and urban development sectors must incorporate groundwater considerations.	Managed Aquifer Recharge should be greatly increased globally.

Table 3.1 contd. ...

...Table 3.1 contd.

Effort Principle	Groundwater Governance: A Global Framework for Action	Groundwater Visibility Initiative	ISMAR9 Policy Declaration
Integrated water management		Effective groundwater management is critical to an integrated water management portfolio that is adaptive and resilient to drought and climate change.	Groundwater must be sustainably managed and protected, within an integrated water resource framework.
Protecting groundwater resources	Incentive frameworks and investment programs foster sustainable, efficient groundwater use and adequate groundwater resources protection.		Halt the chronic depletion of groundwater in aquifers on a global basis. Recognize aquifers and groundwater as critically important, finite, valuable and vulnerable resources.

Groundwater Governance: A Global Framework for Action

In response to emerging concerns over increasingly unsustainable use of groundwater across the globe, the GEF Secretariat requested the project "Groundwater Governance: A Global Framework for Action" (FAO/GEF 2010). The primary partners in this effort were the World Bank, United Nations (UN) Educational, Scientific and Cultural Organization's International Hydrological Programme, the Food and Agriculture Organization of the UN, and the International Association of Hydrogeologists. The project's objective was to "influence political decision making by achieving a significantly increased level of awareness of the paramount importance of sustainable groundwater resources management in averting the impending water crisis" (FAO/GEF 2010). The project was built around three main lines of action. The first line of action was entitled "Building on Existing Knowledge Base and Initiatives". This was meant to focus on drawing lessons and experiences from ongoing and past programs and projects to consolidate and synthesize information on groundwater governance. The second line of action, "Strengthening Partnerships", focused on strengthening partnerships between UN Water and its member agencies and programs, the Consultative Group on International Agricultural Research, and other organizations and national geological surveys with a history of international cooperation on groundwater. The third line of action, "Mainstreaming Groundwater in the GEF Programs and Projects", aimed to elevate the profile of groundwater management in its project portfolio (FAO/GEF 2010).

The project produced three key project documents (available at www.groundwatergovernance.org): (1) *Shared Global Vision for Groundwater Governance in 2030*; (2) *Global Framework for Action*; and (3) *Global Diagnostic on Groundwater Governance*. A fundamental finding of the project is that the world's groundwater resources are suffering due to a lack of effective governance. The Shared Global Vision offers several aspirations for 2030:

- There are appropriate and implemented legal, regulatory and institutional frameworks for groundwater that establish public guardianship and collective responsibility, permanent engagement of stakeholders and beneficial integration with other sectors, including other uses of the subsurface space and its resources.
- All major aquifers are properly assessed, and the resulting information and knowledge are available and shared, making use of up-to-date information and communication techniques.
- Groundwater management plans are prepared and implemented for the priority aquifers.
- Groundwater management agencies, locally, nationally and internationally, are resourced and their key tasks of capacity building,

resource and quality monitoring, and promoting demand management and supply-side measures are secured.
- Incentive frameworks and investment programs foster sustainable, efficient groundwater use and adequate groundwater resources protection (GEF/World Bank 2016).

Workshop Report of the Groundwater Visibility Initiative

The AWRA and NGWA convened 24 water experts from the U.S. and Canada in April 2016 for a Groundwater Visibility Initiative workshop. Its purpose was to find ways to "elevate groundwater's status in the international discourse on water policy, governance, and management by crafting recommendations for action" (GVI 2016). The published findings of the workshop (Alley et al. 2016a, Alley et al. 2016b) are summarized in Table 3.1. The first finding was that governing and managing groundwater requires working with people. People who should be involved with governing and managing include stakeholders, multidisciplinary teams and alliances among multiple governance/management associations. Next, the experts found that data and information are the key for increasing knowledge for water withdrawals and consumptive use. The experts also acknowledged in the third principle that some "secrets" remain, particularly regarding the understanding of climate impacts on quantity and quality, demand for groundwater, and groundwater-surface water interactions. The fourth principle stated, "we need to take care of what we have", meaning that the incorporation of infrastructure rehabilitation and maintenance should be incorporated within planning and investment. Experts also concluded that effective groundwater management is critical to an integrated water management portfolio that is adaptive and resilient to drought and climate change. Finally, experts advised that policies of agriculture, energy, environment, land-use planning, and urban development sectors must incorporate groundwater considerations (GVI 2016).

ISMAR9 Policy Declaration

At the June 2016, ISMAR9 conference in Mexico City, Mexico, participants developed a document for decision-makers and the public to inform, engage, and educate stakeholders on the "critical need" to address depleting groundwater resources before it is too late (ISMAR9 2016). The document consisted of six policy directives. The first directive recognizes aquifers and groundwater as critically important, finite, valuable and vulnerable resources. Second, the document advises to halt the chronic depletion of groundwater in aquifers on a global basis. The document writers also wanted to stress that aquifer systems are unique, need to be well understood, and groundwater should be invisible no more. The next principle advocates for sustainable management and protection of groundwater, within an integrated water resource framework. The document also advocates for the management technique of Managed Aquifer Recharge, saying that it should be greatly increased globally. Finally, the document writers stated that effective groundwater management requires collaboration, robust stakeholder participation and community engagement (ISMAR9 2016). All of these efforts mentioned in this section have suggestions for improved groundwater management and governance.

Four Pillars of Governance

Groundwater governance can be viewed through many lenses, and a useful review of groundwater governance studies and practices is contained in the study by Varady et al. (2016). This identified four governance "pillars" that are integral to the modes and approaches of groundwater governance: (1) institutional setting; (2) availability and access to information and science; (3) robustness of civil society; and (4) economic and regulatory frameworks. Institutional setting depends on which organizations are making decisions and managing groundwater, be they governmental, nongovernmental, and/or private sector agencies or organizations (Varady et al. 2016). Institutional setting can also refer to the level of decision-making—whether that is top-down or bottom-up—and how the governance interacts with other decision-making institutions. For groundwater governance to be effective, the availability and access to science and information about the resource itself is crucial. This includes having access to groundwater monitoring information. Having this information at disposal for all parties involved is particularly crucial for negotiations over water allocation and management (Moench et al. 2003). The third pillar, robustness of civil society, involves stakeholder engagement in decision-making. This is found to be critical to groundwater governance as stakeholder behavior determines how the resource is utilized. Their activities can cause or prevent pollution and overdrafts (Garduño et al. 2010). Finally, regulations regarding groundwater use, quality, and monitoring may determine how economic incentives and the responses to the said incentives may influence how groundwater is used (Varady et al. 2016).

These pillars of groundwater governance share commonalities of the principles espoused within the groundwater governance visibility efforts summarized in Table 3.1. All three efforts also tangentially refer to the first pillar, institutional setting, through calls for stakeholder engagement and inclusion. The second pillar, availability and access to information and science, is included in all three efforts, as is the third pillar, robustness of civil society (labeled in Table 3.1 as stakeholder engagement and inclusion). The fourth pillar, economic and regulatory frameworks, is included within the Groundwater Governance Framework principles and the Groundwater Visibility Initiative. Also, both the Groundwater Visibility Initiative and the ISMAR9 Policy Declaration call for integrated water management. We discuss how these pillars manifest in transboundary groundwater governance dynamics in the next section. The fundamental goal for these studies and efforts is to further the development and implementation of good groundwater governance approaches.

Groundwater Governance in a Transboundary Setting

Transboundary Aquifers in the World

As shown in Fig. 3.1, the International Groundwater Resources Assessment Centre (IGRAC), a UNESCO Category 2 center, has identified 592 transboundary aquifers, underlying almost every nation in the world (IGRAC 2015). Undoubtedly, the list will grow as more are identified. Currently, at least 8% of transboundary aquifers worldwide are stressed due to overexploitation (Wada and Heinrich 2013). As the rate of groundwater pumping has increased substantially in the last 50 years, the aquifer stress of many transboundary aquifers across Europe, Asia and Africa has been increasing (Wada and Heinrich 2013).

The lack of institutional structures for resolving transboundary groundwater problems make matters more difficult. The absence of water allocation laws/regulations, institutions with conflict resolution authority, and available data for resource boundaries, capacities, and conditions is very likely in international aquifers (Blomquist and Ingram 2003).

International Principles and Laws on Transboundary Aquifers

While transboundary surface water management has traditionally involved sovereign states entering into international agreements to maintain sovereignty over actions that may harm or benefit them, this approach has been less successful for managing groundwater (Lopez-Gunn and Jarvis 2009). This is due to many reasons, including the "invisible" nature of groundwater, lack of data collection and monitoring, the large amount of uncertainty from groundwater conceptual models, mismatches in scale, and deeply rooted conflicts in authority, knowledge, and territory. All of these factors lead to a general lack of institutional capacity for groundwater governance and management (Lopez-Gunn and Jarvis 2009).

Conti and Gupta (2016) assessed 12 key international groundwater governance texts and found principles that could be necessary to achieve sustainable groundwater governance (i.e., groundwater's linkages with all water resources, including all groundwater resources; the potential impact of climate change on water resources; and the impact of trade on equitable sharing between regions and protection of groundwater-related ecosystems). The content of these texts has evolved rapidly in the last 25 years, with an increase in the number of principles. Governments could make more progress in including these principles in legally binding groundwater texts (Conti and Gupta 2016).

Three UN water law documents have particular relevance to the development of principles for governance of transboundary aquifers. They are (1) the 1992 UN Economic Commission for Europe Convention on the Protection and Use of Transboundary Watercourses and International Lakes (UNECE Water Convention), (2) the 1997 UN Convention on the Law of the Non-Navigational Uses of Transboundary Watercourses (UN Convention), and (3) the 2008 UN Draft Articles on the Law of Transboundary Aquifers (Draft Articles). The first two documents primarily focus on the management and development of transboundary surface water systems. The Draft Articles, arguably the most substantial effort to develop an international regulatory system for transboundary aquifer systems (Eckstein 2011), are discussed below.

UN General Assembly Resolution A/RES/63/124 on the Law of Transboundary Aquifers

The UN undertook an effort to develop an international regulatory system for transboundary aquifers, which resulted in the UN General Assembly adopting Resolution A/RES/63/124, containing 19 draft articles (UNGA 2009). The UN Draft Articles are modeled largely on the UN Convention, including within its framework the rules of equitable and reasonable utilization and no significant harm (Eckstein 2011). The Draft Articles also include provisions to readily exchange data and information, protect and preserve ecosystems, prevent and minimize any detrimental impacts on recharge and discharge zones, prevent and control pollution, monitor the aquifers or aquifer systems, and give prior notification of planned activities that could cause adverse effects (UNGA 2009).

The Draft Articles include groundwater regardless of whether it is recharged by or discharged to an international watercourse and governs activities that have or are likely to have an impact upon transboundary aquifers within its scope (Yamada 2011). Another noteworthy component of the Draft Articles is the definition of the term "aquifer", which includes both the geological formation that serves as the container of water and the water within the saturated zone of the formation. The definition's inclusion of the geological formation allows regulation of the aquifer's utilization for means such as storage, waste disposal, and carbon sequestration (Yamada 2011).

Article 3 calls for state sovereignty over the portion of a transboundary aquifer located within its territory (UNGA 2009). McCaffrey (2011) argues that this provision requires a fundamental rethinking of how basic principles within the UN Convention, including equitable and reasonable utilization, prevention of significant harm and prior notification of planned measures, would be applicable to the Draft Articles.

Limited progress has been made in ratification of the Draft Articles since their introduction. Since 2008, the Draft Articles have been on the agenda of the UN General Assembly four times (2008, 2011, 2013, and 2016). In 2008 and 2011, the UN General Assembly's Sixth Committee discussed the Draft Articles, subsequently tabling them for consideration at future meetings. The law will be included in the UN General Assembly's provisional agenda of its 74th session (UNGA 2016).

Though the Draft Articles are not in effect, they have already influenced state practice and the development of international law. The Guarani Aquifer Agreement and the 2009 Bamako Declaration for the Iullemeden Aquifer System both reference the Draft Articles (Eckstein and Sindico 2014). In addition, the UNECE Water Convention's Model Provisions reference the Draft Articles for giving guidance in groundwater practices (Eckstein and Sindico 2014).

Partly in response to these developments in international law, policymakers have been increasing their attention to transboundary aquifers through policy and law-making initiatives, several negotiation efforts, and expert dialogues. Countries have signed agreements on the management of transboundary aquifers, including the French-Swiss Genevese aquifer (Convention Genevois 2008), North Africa's Nubian Sandstone aquifer (NSAS 2002) and Northwestern Sahara aquifer (SASS 2002), the Iullemeden Aquifer System of West Africa (Eckstein 2011), the Al-Sag/Al-Disi aquifer between Jordan and Saudi Arabia, and South America's Guarani Aquifer (Guarani Acuerdo 2010). However, only the Genevese aquifer is collaboratively managed. The Al-Sag/Al-Disi aquifer has rudimentary extraction controls, and the two northern Africa aquifers (Nubian Sandstone and Northwestern Sahara) have data sharing agreements (Sanchez et al. 2016). Policymakers are also beginning to consider principles from the Draft Articles into consideration when crafting agreements. The Guarani Aquifer Agreement is the first agreement signed under the influence of the Draft Articles (Villar and Ribeiro 2014). Below, we offer brief descriptions of two of these efforts: cooperation on the Franco-Swiss Genevese Aquifer and the Guarani Aquifer Initiative.

Agreement on the Franco-Swiss Genevese Aquifer

The Agreement on the Franco-Swiss Genevese Aquifer is important due to it being the first treaty dealing exclusively with the management of a transboundary aquifer. Though the French users were initially reluctant to participate in the negotiation of managing the aquifer, the French communes eventually accepted due to information about the relative costs of exploiting alternative sources of water. Another reason that the French communes agreed is that not all of the communes reliant upon the aquifer would be able to benefit from the intervention of the Swiss (Walter 2013).

The agreement came into effect in 1978 and was revised in 2008 (Convention Genevois 2008). The treaty is also the first that manages and specifically allocates the waters of a transboundary aquifer (Eckstein 2011). The agreement addresses groundwater quality, quantity, abstraction and artificial recharge. The agreement also created the joint Genevese Aquifer Management Commission. While the commission is only consultative, its recommendations and technical opinions carry considerable weight (Eckstein 2011). Walter (2013) identifies three key elements that allowed for successful co-management of the Genevese Aquifer: the existence and recognition of problems with the aquifer, aligning political interests in favor of cooperatively resolving the problems, and effective measures devised to resolve the aquifer's problems. Some of those effective measures include allocation of expenses between the countries for the Swiss recharge efforts and the placement of strict withdrawal limits for the French communes (Eckstein 2011). The 2008 revision strengthened the existing relationship, establishing a joint monitoring program for the Swiss and French authorities to conduct within their own respective territory, each controlling water levels and water quality (Allan et al. 2011).

The agreement is also unique in its nature due to the parties arranging it through sub-national, as opposed to international, channels (Eckstein and Eckstein 2005). The Genevese Agreement also holds significance in its balance between state sovereignties and responsibilities, and lack of provisions related directly to sovereign rights to the aquifer (Eckstein 2011). It is considered one of the most comprehensive sources for guidelines for management and allocation of both surface water and groundwater resources (Gander 2014) and is unique in lacking any provisions directly related to nations' sovereign rights to the underlying aquifer or the waters contained within (Eckstein 2011).

Guarani Aquifer Agreement

The Global Environment Facility-funded Guarani Aquifer Program for groundwater resources sustainability and environmental projection was launched in May 2003 by Argentina, Brazil, Paraguay and Uruguay under the supervision of the World Bank, coordination of the Organization of American States, and with support from the International Atomic Energy Authority (Foster et al. 2009). Though the Guarani Aquifer lacks "serious or high-profile transboundary groundwater quality issues" (Sugg et al. 2015), the project aimed to be preventative in character with its anticipation of potential problems due to future expansion of the aquifer's use for public water supply, hydrogeothermal exploitation and irrigation, as well as significant land-use changes that could impact aquifer recharge rates and/or water quality (Foster et al. 2009). The participating countries developed science-based, local-level operational arrangements, such as when Salto, Uruguay and Concordia, Argentina adopted common drilling practices and regulations involving waste disposal and well distances (Sugg et al. 2015).

The countries involved in this project signed an agreement in 2010 (Villar and Ribeiro 2014). Like the Draft Articles, the Guarani Aquifer Agreement features provisions (Articles 2 and 3) that call for sovereignty over the water resources falling in their respective territories (Guarani Acuerdo 2010) that may be viewed as controversial to those who advocate for more communal practices in transboundary water management. Sindico (2011) argues that these provisions within the Guarani Aquifer Agreement are not as disappointing as they may seem, as (1) national sovereignty still must be exercised in accordance with domestic and international legal restraints; (2) national sovereignty is seen as a way for participating countries to combat "hydro-myths" where foreign powers would steal water from the Guarani Aquifer System; and (3) relinquishing sovereignty over the water would also mean relinquishing sovereignty over the geological formations within the aquifer. The 2010 agreement also features obligations to cooperate and dispute resolution mechanisms (Guarani Acuerdo 2010).

The agreement is notable for its proactivity. Though some Guarani Aquifer regional experts suggest that while challenges are mostly localized, urban, and not transboundary at this point, these problems could worsen. Understanding and anticipating problems in advance could be very useful. That said, the agreement is still not fully ratified. Though Argentina and Uruguay have fully ratified the agreement, Brazil has yet to make an internal ratification decision and Paraguay has refused to ratify, due to perceived limits to its national sovereignty (Sugg et al. 2015).

The Importance of Asymmetries Associated with Different Nations/ Governing Bodies

Asymmetries between different nations and governing bodies in how groundwater is managed are bound to exist, including between governing bodies in the Genevese and Guarani aquifers. To effectively address asymmetries with different nations or governing bodies, Feitelson and Haddad (1998) advocate for an open-ended, iterative approach for identifying joint groundwater management structures, due to a paucity of experience in the management of transboundary aquifers and the likelihood that viable options for the management of shared aquifers have not been tried yet.

Addressing asymmetries can also be difficult due to governance principles, conventions, rules, and treaties having differing scopes (particularly at the global and regional-transboundary levels), historical water governance developments are disjointed and still need to be streamlined, and there is poor policy enforcement at the local level (Conti and Gupta 2016). Sometimes, laws governing transboundary groundwater can compete for primacy within a transboundary aquifer, such as between Mexico and the U.S., where Mexico's system primarily focuses on federal allocation, and the U.S. has a state-centric system (Megdal and Scott 2011, Conti and Gupta 2014).

The authors next discuss how the Transboundary Aquifer Assessment Program (TAAP) along the U.S.-Mexico border is consistent with elements of the Draft Articles. One could argue that it is difficult to govern what has not yet been characterized. Therefore, cooperative efforts to assess binational aquifers can be considered a precursor to binational governance and management. Efforts to characterize two aquifers shared by the U.S. state of Arizona and the Mexican state of Sonora are evidence of the importance of establishing a cooperative framework early on and having a binational organization in place to guide collaboration.

Aquifer Assessment along the U.S.-Mexico Border

Transboundary Aquifer Assessment Program

The Transboundary Aquifer Assessment Program (TAAP) started with legislation enacted in the U.S. Public Law 109-448, known as the United States-Mexico Transboundary Aquifer Assessment Act (Act), established a governmental-academic partnership to assess transboundary aquifers (TAAA 2006). The law authorized the Secretary of the Interior (through the U.S. Geological Survey (USGS)) to partner with university-based water resources research institutes for tasks including hydrogeologic characterization, mapping, and modeling for priority transboundary aquifers. The Act specified certain priority aquifers (Fig. 3.2) as initial focal points for the TAAP, based on considering a border aquifer's proximity to densely populated areas, the extent of aquifer utilization, and the transboundary aquifer's susceptibility to contamination (Megdal and

54 Lake Governance

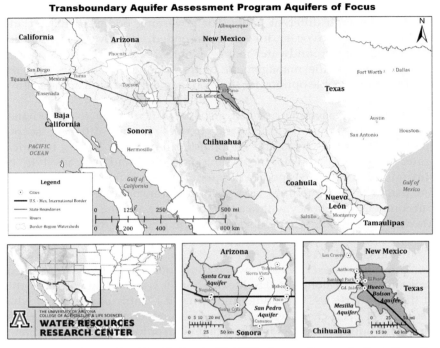

Figure 3.2. Priority aquifers identified by the TAAP. The four aquifers are the Santa Cruz, the San Pedro, the Mesilla, and the Hueco Bolson.

Scott 2011). The Act included a 10-year authorization, which expired late in 2016. It also authorized up to US$50 million, but only US$4 million in appropriations have been approved by the U.S. Congress to date. Despite the limited funding relative to expectations or hopes for funding, the program has produced results as described below.

The International Boundary and Water Commission as the Binational Coordinating Agency

The International Boundary and Water Commission (IBWC) is a binational organization whose mission is "to provide binational solutions to issues that arise during the application of U.S.-Mexico treaties regarding boundary demarcation, national ownership of waters, sanitation, water quality, and flood control in the border region" (IBWC, n.d.). The U.S. and Mexican sections of the IBWC, located within their respective country's state or foreign affairs secretariat, coordinate through established mechanisms. The border area of IBWC responsibility includes several surface water basins. Both countries have had long-standing mandates related to binational surface water and wastewater treatment through the IBWC. Perhaps the IBWC is best known for its authority to make rules through adopting "Minutes" to the 1944 Treaty on the Utilization of Waters of the Colorado and Tijuana Rivers and of the Rio Grande. This process enables flexibility for the IBWC, which in turn can help to secure long-term compliance with the 1944 Treaty (Umoff 2008). Despite the hydrologic interconnectivity of surface water and groundwater, no treaty exists on groundwater management for the shared aquifers shared between Mexico and the U.S. According to one study, Mexico and the U.S. share potentially 36 transboundary aquifers (Sanchez et al. 2016).

The IBWC platform for international water cooperation on binational surface water management and wastewater treatment has played a critical role in implementing TAAP. Although groundwater governance in the U.S. is left to the states, Mexico's governance is centralized at the federal level (Megdal and Scott 2011). Regardless of the limited role of the U.S. government in groundwater governance, international efforts related to water of any kind requires transboundary cooperation. In addition, U.S. legislation cannot require or otherwise bind Mexico to commitments of any kind. Any involvement of Mexico in transboundary aquifer assessment, including agreement to study specific aquifers, is completely at Mexico's prerogative. Therefore, the IBWC served as the ready vehicle for binational discussions of the desirability of and methodology for conducting transboundary aquifer assessment. Given both countries' recognition of the importance of better scientific understanding of shared aquifers, discussions of the principles for a cooperative framework for undertaking aquifer assessment began. As collaborative aquifer assessment was a new undertaking, it took some time to approve the framework, consistently expressed in English and Spanish. The *Joint Report of the Principle Engineers Regarding the Joint Cooperative Process United States-Mexico for the Transboundary Aquifer Assessment Program* was signed by U.S. and Mexican Principal Engineers on 19 August 2009. This Joint Report is notable as it serves as the framework for coordination between Mexico and the U.S. and dialogue for implementing transboundary aquifer studies (IBWC 2009).

The considerable time to develop the three pages of text was a good investment because the Joint Report has guided the program well. The Joint Report acknowledges:

> "The IBWC, United States and Mexican Sections, are aware of the interest on both sides of the border to preserve and understand the aquifers used by both countries, whereby it is considered necessary to develop a team of binational experts

to assess aquifers, exchange data, and if needed, develop new datasets.

Initiatives that include transboundary water resources are traditionally coordinated through the IBWC using the customary binational cooperation process used by both Sections of the Commission. The IBWC, under this joint cooperative process, will provide the framework for coordination of binational assessment activities conducted by U.S. and Mexican agencies, universities, and others participating in the program" (IBWC 2009).

The Joint Report clearly establishes the focus of the effort is "to improve the knowledge base of transboundary aquifers between the United States and Mexico" (IBWC 2009). Assuring the concurrence of both countries of transboundary aquifer assessment activities is paramount. It specifies that binational technical advisory committees will be established for each identified transboundary aquifer and that the IBWC will be the official repository for binational project reports, which will be published in Spanish and English. Responsibilities include an allowance for either of the two countries to propose an aquifer to the IBWC to study. Within the framework of the Joint Report, the IBWC is responsible to determine whether a proposed aquifer study is in the interest of both countries, and to develop a joint program. The IBWC also coordinates with agencies for both countries in defining the scope of the assessment and facilitating concurrence of work plans (IBWC 2009).

Funding provisions state that "each country will be responsible for any costs on projects conducted in its territory, in addition to selecting the participants and consultants to carry out the studies in that country. Each country may contribute to costs for work done in the other country" (IBWC 2009), with the IBWC coordinating any flow of funds across the border.

The Joint Report continues by listing six Principles of the Agreement. Key among the principles are (1) activities should be beneficial to both countries and (2) no provision will limit what each country can do independently within its boundaries. Importantly, the agreement cannot contravene provisions of existing water treaties between the countries. It stipulates that "information generated from these projects is solely for the purpose of expanding knowledge of the aquifers and should not be used by one country to require that the other country modify its water management and use" (IBWC 2009).

The final section on communication and use of information specifies that reports will be posted on each section's web site after approval within the framework of the IBWC. Information shall be considered official data and shared without restrictions with stakeholders who participate in the joint projects (IBWC 2009).

The Joint Report provides a concise and elegant foundation for accomplishing the kind of cooperation envisioned by the Draft Articles discussed above. Specifically, they align with Article 7's general obligation to cooperate (through the establishment of joint mechanisms of cooperation) and Article 8's principle of regular exchange of data and information (Megdal 2017).

Implementing Aquifer Assessment for the San Pedro and Santa Cruz Aquifers

The Joint Report established what we refer to as the "Cooperative Framework" for aquifer assessment for agreed-upon transboundary aquifers. Two aquifers along the portion of the border shared by Arizona and Sonora (Fig. 3.3) were specified for assessment—the Santa Cruz and San Pedro. The partnership between governmental agencies and universities has been instrumental for the assessment of these two aquifers. On the U.S. side, the Act specified that 50% of appropriated U.S.

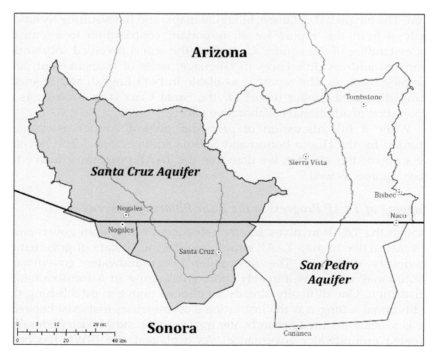

Figure 3.3. The Santa Cruz and San Pedro Basin Aquifers shared between the U.S. state of Arizona and the Mexican state of Sonora.

funds would be allocated to the USGS, with the other 50% being directed to federally authorized water resources research institutes/centers in participating states. The Act did not legislate a split across participating states or priority aquifers. The Act also allows for the U.S. water institutes/centers to subcontract with partners, including those in Mexico (though a match in funds is required from Mexico for subcontracting with a Mexican agency within a partnership).

Under the Cooperative Framework, the Universidad de Sonora, the Comisión Nacional del Agua, the USGS, and the University of Arizona have partnered on aquifer studies for the transboundary San Pedro and Santa Cruz River Basin Aquifers. Sustained and substantial efforts culminated in release of the *Binational Study of the Transboundary San Pedro Aquifer* in December 2016 (Callegary et al. 2016). The report marks the first time that researchers from both countries have collaborated to collect and analyze data and jointly prepare binational maps of the San Pedro River aquifer. It covers various topics about the San Pedro Aquifer, including information about its physical geography, hydrography, hydrometeorology, geology, geophysics, hydrogeology, and hydrogeochemistry. The report also identifies data gaps and information to be updated for subsequent phases and provides a set of recommendations from the binational working team. The binational database, bilingual maps, and the resulting technical analysis from this report are an important contribution to scientists' understanding of the aquifer. Compiling the report involved substantial efforts to address differences in language, units of measurement, and mapping systems. The report is available in both English and Spanish. Completion of a similar report for the Santa Cruz is in process, as is exploration of additional collaborative efforts.

While a full discussion of past and present aquifer assessment activities for the Hueco Bolson and Mesilla aquifers (Fig. 3.2) is beyond the scope of this chapter, we note that the TAAP program is active for these aquifers as well.

Mapping of TAAP Progress to the Four Pillars of Governance

Though the TAAP involves aquifer assessment rather than governance, it is instructive to map TAAP efforts into the four pillars of governance previously discussed. The first pillar of groundwater governance, institutional setting, is arguably more challenging in a transboundary jurisdiction. One difficulty already mentioned above in establishing the institutional setting was the institutional asymmetries that exist between the U.S. (state-oriented groundwater governance) and Mexico (federally-oriented groundwater governance). As explained, the IBWC has been integral to coordinating among the various institutions involved in the TAAP. In fact, it can be argued that the existence of a well-established border water agency facilitated the development of the Joint Report, which provided the foundational Cooperative Framework.

The TAAP efforts are increasing the availability and access to information and science through efforts such as the ongoing hydrological and climatological studies on the Santa Cruz Basin (including the preparation of the *Binational Study of the Transboundary Santa Cruz Aquifer*) and the *Binational Study of the Transboundary San Pedro Aquifer*. The binational database, bilingual maps, and the resulting technical analysis from the study are publicly available in English and Spanish (https://ibwc.gov/Files/San_Pedro_River_Binational%20Report_013116.pdf).

Stakeholder engagement in the TAAP planning process (robustness of civil society) has been sporadic to date. A 2009 workshop to define focal areas for the San Pedro and Santa Cruz aquifers engaged regional and even international experts. TAAP investigators made efforts to include information about the status of data and modeling, to visit sites to understand the characteristics of the aquifer regions, to develop relationships and communication channels important for program implementation, and to disseminate information about the TAAP purpose and progress (Megdal and Scott 2011, https://wrrc.arizona.edu/TAAP). Additionally, the U.S. Section of IBWC has held Citizens' Forums on the U.S. side of the border in efforts to increase stakeholder engagement (Varady et al. 2013). Stakeholder engagement efforts in the TAAP continue, including a forum in June 2017 (see TAAP-A/S 2017).

Regulations for groundwater use, quality, and monitoring between countries have yet to be harmonized. The Mexican portions are subject to federal regulation, while the U.S. portions of the aquifers are bound to rules on water quality and quantity set by the state of Arizona and the federal government.

Importantly, the TAAP is explicitly limited to aquifer assessment, not governance and management matters. However, as suggested above, it is difficult to govern and manage something that is not well understood or characterized. The aquifer assessment activities to date contribute to enhanced understanding of the binational aquifers studied and can be the basis for further assessment and/or dialogue on issues related to governance and management.

Concluding Remarks

This chapter has focused on the emerging principles of transboundary groundwater governance. Developing approaches to groundwater governance in a transboundary setting is complex and will be iterative. Agreement on principles is developing, but more experience with actual transboundary groundwater governance is needed to assess the application of these principles. The three initiatives to increase the visibility of groundwater governance can assist in assessing options. Greater exchange of experiences across regions and types of waters is likely to assist in the identification of good practices for transboundary water governance.

While not going as far as actual transboundary groundwater governance and management, the aquifer assessment experience along the U.S.-Mexico border suggests that institutional readiness to work across borders is very helpful in establishing the mutually respectful framework, including communication channels, needed for transboundary cooperation. Though developed for aquifer assessment, the Cooperative Framework articulated within the Joint Report contains elements that are readily transferable to assessment of other waters.

The cases and efforts discussed in this chapter underscore the importance of continuing to share lessons learned and further refine principles and best practices. Programs and initiatives will continue to increase the visibility and understanding of the complications, along with opportunities, associated with governing groundwater in a transboundary setting.

Acknowledgements

The authors wish to thank Elia Tapia for her review of their chapter and her mapping assistance. Funding has been provided through the U.S. Geological Survey budget to the University of Arizona Water Resources Research Center through multiple grants. In addition, University of Arizona Technology Research and Initiative Funding has supported TAAP efforts in part. Author Sharon B. Megdal acknowledges her appreciation for the opportunity to work on TAAP since testifying in 2006 in the U.S. House of Representatives on the authorizing legislation. Analysis contained in this article is based on her first-hand knowledge and experience and can therefore be provided without the usual citations and references.

References

9th International Symposium on Managed Aquifer Recharge (ISMAR9). (2016). ISMAR9 Call to Action: Sustainable Groundwater Management Policy Directives. June 2016, Mexico City, Mexico. http://www.ismar9.org/Doc/SUSTAINABLEDIRECTIVES.pdf. Accessed 4 November 2016.

Allan, A., Loures, F. and Tignino, M. (2011). The role and relevance of the Draft Articles on the Law of Transboundary Aquifers in the European context. *Journal for European Environmental & Planning Law*, 8(3): 231–251.

Alley, W.M., Beutler, L., Campana, M.E., Megdal, S.B. and Tracy, J.C. (2016a). Groundwater visibility: the missing link. *Groundwater*, 54(6): 758–761.

Alley, W.M., Beutler, L., Campana, M.E., Megdal, S.B. and Tracy, J.C. (2016b). The groundwater Visibility Initiative: integrating groundwater and surface water management. Report of Workshop held in Denver, Colorado, April 28, 2016. National Groundwater Association and American Water Resources Association. 20 August, 2016.

Blomquist, W. and Ingram, H.M. (2003). Boundaries seen and unseen: resolving transboundary groundwater problems. *Water International*, 28(2): 162–169. DOI: 10.1080/02508060308 691681.

Braune, E. and Adams, S. (2013). Regional diagnosis for the sub-Saharan Africa region. Groundwater governance—a global framework for action. GEF, UNESCO-IHP, FAO, World Bank and IAH. www.groundwatergovernance.org. Accessed 3 February 2017.

Callegary, J.B., Minjárez Sosa, I., Tapia Villaseñor, E.M., dos Santos, P., Monreal Saavedra, R., Grijalva Noriega, F.J. Huth, A.K., Gray, F., Scott, C.A., Megdal, S.B., Oroz Ramos, L.A., Rangel Medina, M. and Leenhouts, J.M. (2016). Binational Study of the Transboundary San Pedro Aquifer. *International Boundary and Water Commission*, 170 p. http://ibwc.gov/Files/San_Pedro_River_Binational%20Report_013116.pdf. Accessed 15 December 2016.

Condon, L.E. and Maxwell, R.M. (2015). Evaluating the relationship between topography and groundwater using outputs from a continental-scale integrated hydrology model. *Water Resources Research*, 51: 6602–6621. DOI:10.1002/2014WR016774.

Conti, K.I. and Gupta, J. (2014). Protected by pluralism? Grappling with multiple legal frameworks in groundwater governance. *Current Opinion in Environmental Sustainability*, 11: 39–47.

Conti, K.I. and Gupta, J. (2016). Global governance principles for the sustainable development of groundwater resources. *International Environmental Agreements: Politics, Law and Economics*, 16(6): 849–871.

Convention Genevois. (2008). Convention relative a la protection, a l'utilisation, a la realimentation et au suivi de la Nappe Souterraine Franco-Suisse du Genevois [online]. Geneva, 18 December 2007; in force on 1 January 2008. http://internationalwaterlaw.org/documents/regionaldocs/2008Franko-Swiss-Aquifer.pdf. Accessed 14 November 2016.

De Chaisemartin, M., Varady, R.G., Megdal, S.B., Conti, K.I., van der Gun, J., Merla, A., Nijsten, G.J. and Scheibler, F. (2017). Addressing the groundwater governance challenge. pp. 205–227. *In*: Karar, E. (ed.). Freshwater Governance for the 21st Century. Springer International Publishing, Cham, Switzerland.

Döll, P., Schmeid, H.M., Schuh, C., Portman, F.T. and Eicker, A. (2014). Global-scale assessment of groundwater depletion and related groundwater abstractions: combining hydrological modeling with information from well observations and GRACE satellites. *Water Resources Research*, 50(7): 5698–5720.

Eckstein, G.E. (2011). Managing buried treasure across frontiers: the international Law of Transboundary Aquifers. *Water International*, 36(5): 573–583.

Eckstein, G.E. and Sindico, F. (2014). The Law of Transboundary Aquifers: many ways of going forward, but only one way of standing still. *Review of European, Comparative & International Environmental Law*, 23(1): 32–42.

Eckstein, Y. and Eckstein, G.E. (2005). Transboundary aquifers: conceptual models for development of international law. *Groundwater*, 43(5): 679–690.

Food and Agriculture Organization of the United Nations/Global Environment Facility (FAO/GEF). (2010). Groundwater Governance: A Global Framework for Country Action. http://www.groundwatergovernance.org/fileadmin/user_upload/gwg/documents/GWG%20prodoc%20update%20(4).pdf. Accessed 3 November 2016.

Famiglietti, J.S. (2014). The global groundwater crisis. *Nature Climate Change*, 4: 945–948.

Feitelson, E. and Haddad, M. (1998). A stepwise open-ended approach to the identification of joint management structures for shared aquifers. *Water International*, 23(4): 227–237.

Foster, S., Hirata, R., Vidal, A., Schmidt, G. and Garduno, H. (2009). The Guarani Aquifer Initiative - Towards Realistic Groundwater Management in a Transboundary Context.

Sustainable Groundwater Management: Lessons from Practice. The World Bank, Global Water Partnership Program. http://siteresources.world bank.org/INTWRD/Resources/GWMATE_English_CP9.pdf. Accessed 9 September 2012.

Gander, M.J. (2014). International water law and supporting water management principles in the development of a model transboundary agreement between riparians in international river basins. *Water International*, 39(3): 315–332.

Garduño, H., van Steenbergen, F. and Foster, S. (2010). Stakeholder participation in groundwater management. Available online: http://siteresources.worldbank.org/EXTWAT/Resources/4602122-1210186362590/ GWM_Briefing_6new.pdf. Accessed 6 November 2016.

Giordano, M. and Villholth, K.G. (eds.). (2007). The Agricultural Groundwater Revolution: Opportunities and Threats to Development (Vol. 3). CABI, Oxford, UK.

Giordano, M. (2009). Global groundwater? Issues and solutions. *Annual Review of Environment and Resources*, 34: 153–178.

Global Environment Facility/World Bank (GEF/World Bank). (2016). Global Framework for Action to achieve the Vision on Groundwater Governance. http://www.groundwatergovernance.org/fileadmin/user_upload/groundwatergovernance/docs/GWG_FRAMEWORK_EN.pdf. Accessed 4 November 2016.

Groundwater Visibility Initiative (GVI). (2016). The Groundwater Visibility Initiative: Integrating Groundwater and Surface Water Management. Report of Workshop held in Denver, Colorado, April 28, 2016. https://wrrc.arizona.edu/sites/wrrc.arizona.edu/files/GVI-Workshop-Main-Report.pdf. Accessed 4 November 2016.

Guarani Acuerdo. (2010). Acuerdo sobre el Acuífero Guarani. San Juan, Argentina, 2 August. http://www.internationalwaterlaw.org/documents/regionaldocs/Guarani_Aquifer_Agreement-Spanish.pdf. Accessed 14 November 2016.

IBWC-CILA (IBWC). (2009). Joint Report of the Principal Engineers Regarding the Joint Cooperative Process United States-Mexico for the Transboundary Aquifer Assessment Program: 11p., August 19, 2009.

IBWC-CILA (IBWC). N.d. Welcome. http://www.ibwc.gov/home.html. Accessed 27 January 2017.

International Groundwater Resources Assessment Centre (IGRAC). (2015). Transboundary Aquifers of the World. Special Edition for the 7th World Water Forum 2015. https://www.un-igrac.org/sites/default/files/resources/files/TBAmap_2015.pdf. Accessed 10 November 2016.

Jarvis, W.T., Giordano, M., Puri, S., Matsumoto, K. and Wolf, A.T. (2005). International borders, ground water flow, and hydroschizophrenia. *Ground Water*, 43(5): 764–770.

Jarvis, W.T. (2008). Strategies for groundwater resources conflict resolution and management. *In*: Darnault, C.J.G. (ed.). Overexploitation and Contamination of Shared Groundwater Resources. Springer Netherlands, Dordrecht, the Netherlands.

Jarvis, W.T. (2010). Water wars, war of the well, and guerilla well-fare. *Groundwater*, 48(3): 346–350.

Knüppe, K. (2011). The challenges facing sustainable and adaptive groundwater management in South Africa. *Water SA*, 37(1): 67–79.

Lopez-Gunn, E. and Jarvis, W.T. (2009). Groundwater governance and the Law of the Hidden Sea. *Water Policy*, 11: 742–762.

McCaffrey, S.C. (1998). The UN convention on the law of the non-navigational uses of international watercourses: prospects and pitfalls. pp. 17–28. World Bank Technical Paper. Washington, D.C., World Bank, Washington, USA.

McCaffrey, S.C. (2011). The International Law Commission's flawed draft articles on the law of transboundary aquifers: the way forward. *Water International*, 36(5): 566–572.

Megdal, S.B. (2017). The cooperative framework for the transboundary aquifer assessment program: A model for collaborative transborder studies. *Arizona Water Resource*. Summer. https://wrrc.arizona.edu/sites/wrrc.arizona.edu/files/pub-pol-rev-summer-2017_0.pdf.

Megdal, S.B. and Scott, C.A. (2011). The importance of institutional asymmetries to the development of binational aquifer assessment programs: the Arizona-Sonora Experience. *Water,* 3: 949–963. DOI: 10.3390/w3030949.

Megdal, S.B., Gerlak, A.K., Varady, R.G. and Huang, L.Y. (2015). Groundwater governance in the United States: common priorities and challenges. *Groundwater,* 53(5): 677–684.

Moench, M., Burke, J.J. and Moench, Y. (2003). Rethinking the Approach to Groundwater and Food Security. Food & Agriculture Organization, Rome, Italy.

Nubian Sandstone Aquifer System (NSAS). (2002). Programme for the development of a regional strategy for the utilisation of the Nubian Sandstone Aquifer System (NSAS). Terms of reference for the monitoring and exchange of groundwater information of the Nubian Sandstone Aquifer System. Tripoli, 5 October 2000. http://www.fao.org/docrep/008/y5739e/y5739e05.htm. Accessed 14 November 2016.

Puri, S. (ed.). (2001). Internationally shared aquifer resources management, their significance and sustainable management, a framework document, UNESCO-IHP, Series on Groundwater 1. http://unesdoc.unesco.org/images/0012/001243/124386e.pdf. Accessed 3 February 2017.

Puri, S. and Aureli, A. (2005). Transboundary aquifers: a global program to assess, evaluate, and develop policy. *Ground Water,* 43(5): 661–668.

Sahara Aquifer System (SASS). (2002). Establishment of a consultation mechanism for the Northwestern Sahara Aquifer System (SASS). Rome, 19–20 December; endorsed 6 January 2003 (Algeria), 15 February 2003 (Tunisia), 23 February 2003 (Libya). http://www.fao.org/docrep/008/y5739e/y5739e05.htm. Accessed 14 November 2016.

Sanchez, R., Lopez, V. and Eckstein, G. (2016). Identifying and characterizing transboundary aquifers along the Mexico-US border: an initial assessment. *Journal of Hydrology,* 535: 101–119. DOI: 10.1016/j.jhydrol.2016.01.070.

Sindico, F. (2011). The Guarani Aquifer System and the International Law of Transboundary Aquifers. *International Community Law Review,* 13: 255–272.

Sugg, Z.P., Varady, R.G., Gerlak, A.K. and de Grenade, R. (2015). Transboundary groundwater governance in the Guarani Aquifer System: reflections from a survey of global and regional experts. *Water International,* 40(3): 377–400.

Taylor, R.G., Scanlon, B., Döll, P., Rodell, M., Van Beek, R., Wada, Y., Longuevergne, L., Leblanc, M., Famiglietti, J.S., Edmunds, M., Konikow, L., Green, T.R., Chen, J., Taniguchi, M., Bierkens, M.F.P., MacDonald, A., Fan, Y., Maxwell, R.M., Yechieli, Y., Gurdak, J.J., Allen, D.M., Shamsuddha, M., Hiscock, K., Yeh, P.J.F., Holman, I. and Treidel, H. (2013). Ground water and climate change. *Nature Climate Change,* 3(4): 322–329.

Theesfeld, I. (2010). Institutional challenges for national groundwater governance: policies and issues. *Groundwater,* 48(1): 131–142.

Transboundary Aquifer Assessment Program-Arizona/Sonora (TAAP-A/S). 2017 TAAP-A/S Bulletin on the Binational Study of the Transboundary Aquifer. https://wrrc.arizona.edu/sites/wrrc.arizona.edu/files/pdfs/TAAP-bulletin-2017.pdf. Accessed 21 August 2017.

Umoff, A.A. (2008). Analysis of the 1944 US-Mexico water treaty: its past, present, and future. *Environs: Environmental Law and Policy Journal.,* 32: 71–98.

United Nations General Assembly (UNGA) (2009). 63/124. The law of transboundary aquifers. Sixty-third session. Agenda item 75. 15 January 2009.

UNGA. (2016). Resolution adopted by the General Assembly: 71/150. The law of transboundary aquifers. 13 December 2016.

United States-Mexico Transboundary Aquifer Assessment Act (TAAA). (2006). Public Law 109–448. United States Congress: Washington, DC, USA, December 2006. https://www.gpo.gov/fdsys/pkg/PLAW-109publ448/pdf/PLAW-109publ448.pdf. Accessed 2 February 2017.

Varady, R.G., Salmón Castelo, R. and Eden, S. (2013). Key issues, institutions, and strategies for managing transboundary water resources in the Arizona-Mexico border region. pp. 35–50. *In*: Megdal, S.B., Varady, R.G. and Eden, S. (eds.). Shared Borders, Shared

Waters: Israeli-Palestinian and Colorado River Basin Water Challenges. CRC Press, Boca Raton, Florida, USA.

Varady, R.G., Zuniga-Teran, A.A., Gerlak, A.K. and Megdal, S.B. (2016). Modes and approaches of groundwater governance: a survey of lessons learned from selected cases across the globe. *Water*, 8(10): 417. DOI:10.3390/w8100417.

Villar, P.C. and Ribeiro, W.C. (2014). The agreement on the Guarani Aquifer: cooperation without conflict. pp. 69–76. *In*: Grafton, R.Q., Wyrwoll, P., White, C. and Allendes, D. (eds.). Global Water: Issues and Insights. Australian National University Press, Canberra, Australia.

Wada, Y. and Heinrich, L. (2013). Assessment of transboundary aquifers of the world-vulnerability arising from human water use. *Environmental Research Letters*, 8: 2. DOI: 10.1088/1748-9326/8/2/024003.

Wada, Y., Wisser, D. and Bierkens, M.F.P. (2014). Global modeling of withdrawal, allocation and consumptive use of surface water and groundwater resources. *Earth System Dynamics*, 5: 15–40.

Walter, M. (2013). The roles of knowledge in the emergence of co-management initiatives for transboundary groundwaters: the case of the Génévois Aquiver. pp. 292–316. *In*: Grover, V.I. and Krantzberg, G. (eds.). Water Co-Management. CRC Press, Boca Raton, Florida, USA.

Yamada, C. (2011). Codification of the Law of Transboundary Aquifers (Groundwaters) by the United Nations. *Water International*, 36(5): 557–565.

Section 2
Case Studies

CHAPTER 4

Transboundary Water Governance in the Great Lakes Region*

Victoria Pebbles

Introduction

The Great Lakes basin lies within United States (U.S.) and Canada and is the largest freshwater resource in the world, containing nearly 20 percent of the earth's fresh surface water. The basin faces many environmental threats, such as persistent, bioaccumulative and toxic pollutants; invasive species; aging water and sewer infrastructure; excessive nutrient loading; and habitat degradation and loss.

This chapter examines institutional and policy frameworks that govern transboundary water management in the Great Lakes. The Great Lakes region is a useful case study because the region boasts a longstanding history of bilateral environmental cooperation and yet the water management challenges continually evolve. An examination of historic and current institutions and policy arrangements, and associated water management successes and opportunities in the Great Lakes region will shed light on gaps and opportunities for future water management efforts in the Great Lakes region as well as other large bilateral and multilateral aquatic ecosystems.

Program Manager, Great Lakes Commission, 2805 S. Industrial Hwy. Suite 100, Ann Arbor, MI USA 48104.
* This chapter is adapted from Pebbles, V. (2014). Incorporating climate change adaptation into transboundary ecosystem management in the Great Lakes Basin. pp. 197–216. In: Sanchez, J.C. and Roberts, J. (eds.). Transboundary Water Governance: Adaptation to Climate Change. IUCN. Gland, Switzerland. ISBN 978-2-8317-1660-2.

Overview of the Great Lakes Hydrology and Human Development

Natural History and Hydrology

The Great Lakes basin formed about 10,000 years ago at the end of the last ice age as the retreating ice sheet scoured the land, creating ridges valleys and depressions: the largest of these depressions eventually became the Great Lakes. Lakes Superior, Huron, Michigan, Erie, and Ontario, and their connecting channels make up the Great Lakes today[1] (Fig. 4.1). The Great Lakes basin—the watershed that drains into these five lakes and their connecting channels (including Lake St. Clair)—covers 295,000 square miles (76,405,000 hectares). Of that, 94,000 square miles (24,346,000 hectares) is water and 201,000 square miles (52,059,000 hectares) is the surrounding watershed (see light blue on Fig. 4.1) (US EPA and EC 1995).

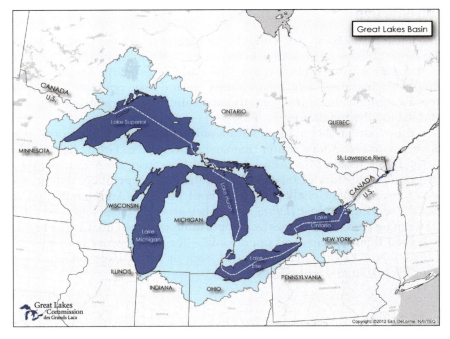

Figure 4.1. Great Lakes Basin. Source: Great Lakes Commission, 2013.

---

[1] Lakes Michigan and Huron are hydrologically connected and considered by scientists as a single lake basin, but they are managed as separate lakes.

Human Settlement and Hydrologic Alterations

The earliest settlers in the Great Lakes region were Native American Indian Tribes. Europeans explorers began scouting the region in the early 1600s and settlements followed thereafter. It wasn't until the 19th century, however, that the region began to experience drastic ecosystem changes due to commercial logging and fishing, industrialization, agricultural intensification and expanding urbanization. The region's hydrology has been altered by navigation locks across the entire system. This includes the Locks at Sault Ste. Marie, Mich. (Soo Locks) that allow passage between Lake Superior and Lakes Michigan and Huron; the Welland Canal which created a navigable bypass around Niagara Falls (between Lakes Erie and Ontario); and the St. Lawrence Seaway locks—actually a series of seven locks that allow ships access to the Atlantic Ocean.

Diversions into and out of the Great Lakes basin also alter its hydrology. Most famous is the Chicago Diversion, completed in 1900 which takes water from Lake Michigan to serve the Chicago metropolitan area and discharges it through the Chicago Sanitary and Ship Canal into the Mississippi River basin.

Modern-Era Environmental Stressors and Uses

Access to freshwater fueled the regions' economic growth. By the early 20th century the Great Lakes region was a hotbed of industrialization and manufacturing for the steel, automobile, pulp and paper and chemical industries. During the same time, the use of chemical and nutrient supports for agriculture also intensified. By-products and wastes, from cities, factories and farms were released directly into the water, land, and air with little attention to the consequences. Coastal development and the dam construction also altered the basin's hydrology and degraded ecosystem functions.

There are approximately 48 million people living in the Great Lakes basin that rely on the Great Lakes for their drinking water. Great Lakes water is also used for industrial uses; agriculture (livestock and irrigation); thermo-electric power (fossil fuel and nuclear); and hydroelectric power generation. Water-based economic activities including commercial shipping, recreational boating, and tourism also rely on Great Lakes water.

Legal, Policy, and Institutional Frameworks for Transboundary Water Management

The Great Lakes basin is part of a shared U.S.-Canada boundary that includes the entire 8,900-km (5,500 mi) border stretching from shared waters between the province of British Columbia, Canada and the U.S.

70 *Lake Governance*

state of Washington, across the continent headwaters of the Gulf of Maine.[2] Governance over the waters and related natural resources of the Great Lakes basin is shared among two federal governments, eight U.S. states (Illinois, Indiana, Michigan, Minnesota, New York, Ohio, Pennsylvania and Wisconsin), two Canadian provinces (Ontario and Quebec), several regional institutions, more than 100 Native American/First Nation authorities, and thousands of local units of government (Hildebrand et al. 2002). A broad range of non-governmental entities, including citizen-based environmental organizations, business associations, industry coalitions, and academic institutions actively and regularly engages these institutions to effectuate Great Lakes governance (Fig. 4.3). Neither the U.S. nor Canada is signatory to the 1997 United Nations Convention on the Law of the Non-Navigational Uses of International Watercourses (United Nations Treaty Collection at (http://treaties.un.org/)).

Water resource management in the U.S. and Canada was governed traditionally by riparian rights common law (originally based on English common law) and associated statutes. Under a riparian rights legal system, water use rights are tied to ownership of (or other legal access to) the land through or under which the water flows. On the U.S. side, adjudication of riparian rights common law has established a legal "reasonable use" doctrine that obliges water users not to cause harm to other users. Similar principles have been incorporated into the statute on the Canadian side of the basin. The practical effect of this legal regime is that the vast majority of water uses in the basin are not regulated.

Binational Policies and Institutions at the Federal Level

The Boundary Waters Treaty and the International Joint Commission

Formal transboundary governance of the Great Lakes began with the 1909 Boundary Waters Treaty (Fig. 4.3) between the United States (U.S.) and Canada which established the International Joint Commission (IJC) to prevent and settle disputes over the boundary waters between the two countries and to approve and manage structures that affect levels and flows in the boundary waters. The IJC is comprised of six Commissioners: three are appointed by the President of the United States and three are appointed by the Prime Minister of Canada. One of those appointees acts as chair of their respective country's delegation on the IJC.

In addition to topic-specific task forces, which generally operate for a certain period, the IJC also has standing advisory boards and regulatory boards that support its work. The IJC also has three Boards of Control

---

[2] The Boundary Waters Treaty also governs management of shared waters between the Alaska (U.S.) and Yukon (Canadian) border.

which have authority to manage structures that affect levels and flows in the boundary waters for hydropower navigation purposes: the Lake Superior Board of Control, the International Niagara Board of Control and the International St. Lawrence River Board of Control. Lakes Erie, Huron and Michigan are indirectly controlled through decisions of the Lake Superior and the International Niagara Boards of Control (Thurber, n.d.). Historically, this control system has been managed to satisfy shipping and hydropower production needs over other economic or ecological needs—an approach which has been challenged over the past decade.

The Great Lakes Water Quality Agreement

The IJC's role in the Great Lakes expanded significantly with the 1972 Great Lakes Water Quality Agreement (GLWQA), a bilateral Executive Agreement between the U.S. and Canada. (As signatories to the GLWQA, the governments of the U.S. and Canada are referred to as "the Parties".) The GLWQA heralded a more ecosystem-based approach by committing both countries to restore and maintain the chemical, physical and biological integrity of the waters of the Great Lakes. The GLWQA also established a separate Great Lakes Regional Office of the IJC to coordinate and oversee implementation of the GLWQA. Subsequent Amendments to the GLWQA in 1978, 1983 and 1987 further strengthened the ecosystem focus. Although the GLWQA is a bilateral Executive Agreement between the two federal governments and therefore does not have treaty status, its goals and objectives have been incorporated into federal, state and provincial laws and policies on both sides of the border (Hildebrand et al. 2002).

The GLWQA was most recently amended in 2012. These latest amendments include 10 annexes to address some of the most pressing transboundary water issues facing the Great Lakes in the 21st century: cleanup of legacy hot spots; nearshore lake management; emerging chemicals, nutrients; vessels; invasive species; habitat; groundwater; climate change; and science. The 2012 amendments also formally established bilateral institutional arrangements that provide structure and process to engage the Parties on a regular basis. A binational Great Lakes Executive Committee (GLEC) led by U.S. EPA and Environment Canada representing the Parties, along with representatives of the eight

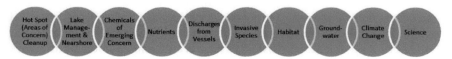

Figure 4.2. Ten Annexes of the 2012 Great Lakes Water Quality Agreement.

U.S. states and two Canadian provinces in the Great Lakes basin was created, replacing an ad-hoc Binational Executive Committee that had been in operation for more than a decade but without formal authority. The 2012 Amendments also formally recognized and established a process for the Parties to "convene a Great Lakes Summit in conjunction with the Great Lakes Public Forum to promote coordination among the Parties, the [International Joint] Commission and other binational and international governmental organizations, and increase their effectiveness in managing the resources of the Great Lakes" (GLWQA 2012 Protocol).

Figure 4.3. Great Lakes Governance Timeline.

State-Provincial and Interstate Policies and Institutions

Several laws, institutions and policies governing transboundary water cooperation at the U.S. state and Canadian province levels are described below.

The Great Lakes-St. Lawrence River Basin Sustainable Water Resources Agreement (Water Resources Agreement) responded to the need for more a robust framework for managing regional water resources arose following a 1999 incident when a pre-existing regional agreement, known as the Great Lakes Charter, proved inadequate to prevent a single jurisdiction from permitting a bulk water export out of the Great Lakes (Annin 2006). The Water Resources Agreement was signed by the governors of the eight U.S. Great Lakes states and premiers of the two Canadian Great Lakes provinces in 2005 and came into force in 2008 marking a new era in transboundary water governance. Similar to the Great Lakes Basin Compact of 1955 that established the Great Lakes Commission, the Water Resources Agreement, each jurisdiction passed their own enabling legislation that affirmed adherence to the Water Resources Agreement. Substantively, the Water Resources Agreement established a new water management framework for managing water diversions and withdrawals. This focus on water quantity complemented the water quality focus of the GLWQA.

The Water Resources Agreement established a new institution to coordinate, and monitor implementation of that policy: the Great Lakes-St. Lawrence River Water Resources Regional Body (Regional Body). Creation of the Regional Body would seem a logical and necessary step if there was not already a rich institutional framework in place, including an existing legally-based interstate compact agency—the Great Lakes Commission—whose members include the same 10 jurisdictions that are party to the Water Resources Agreement. In 2005 when the Water Resources Agreement was signed, the Great Lakes Commission had

Table 4.1. The Great Lakes-St. Lawrence River Basin Sustainable Water Resources Agreement.

- bans new diversions of water from the Basin, with limited exceptions
- sets a consistent standard to review proposed uses "to prevent significant adverse impacts from water withdrawals and losses on the basin ecosystem and its watersheds"
- supports collection and sharing of technical data and information
- requires assessment of cumulative impacts
- requires establishment of water conservation and efficiency programs

been operating for more than 50 years[3] under the authority the Great Lakes Basin Compact, which established the Great Lakes Commission "to promote the orderly, integrated and comprehensive development, use, and conservation of the water resources of the Great Lakes Basin" (Great Lakes Basin Compact 1955). The Great Lakes Commission's broad mandate and longstanding institutional ties to Canada's Great Lakes provinces was unarguably sufficient to accommodate the new directives established under the Water Resources Agreement of 2005.

The Water Resources Agreement is a non-binding good faith agreement; as such, there are no formal mechanisms in the Agreement itself that can be used to force compliance between the two countries, but enforcement mechanisms have been established within each country. Importantly, at the same time the states and provinces signed the Water Resources Agreement, the eight U.S. governors signed a complementary interstate compact [the Great Lakes-St. Lawrence River Basin Water Resources Compact (Water Resources Compact)], which mirrors the requirements of the Water Resources Agreement and provides a legally binding mechanism to ensure compliance among the eight U.S. states that are party to the Water Resources Agreement (U.S. Public Law 110–342; Hall 2010). On the Canadian side, enforcement mechanisms are built into each of the province's implementing

---

[3] The Great Lakes Commission was established by the Great Lakes Compact of 1955 and is implemented through enabling legislation subsequently passed by each of the eight U.S. states that are members of the Great Lakes Commission. The Great Lakes Commission received consent by the U.S. Congress in 1968.

legislation; either of the provinces could use the Canadian justice system to force compliance with their sister province.

Native American Tribes/First Nations are not signatories to the Water Resources Agreement; however, the document sets forth that states and provinces must consult with Tribes/First Nations, in their review of water use or diversion proposals and also calls on the states and provinces to "seek to establish mutually agreed upon mechanisms or processes to facilitate dialogue with, and input from, First Nations and federally recognized tribes."[4]

Notwithstanding the added institutional complexity and risks of redundancy, the Water Resources Agreement responded to the need for a more robust legal and policy framework to manage Great Lakes water resources. The June 2016 decision to allow the City of Waukesha, Wisconsin to divert Great Lakes water tested the Water Resources Agreement (and associated Water Resources Compact) to scientifically assess the merits of and make a firm decision in response to a water diversion application. However, not all stakeholders were satisfied that the standards set by the Water Resource Compact were met and the jury is still out on whether the Water Resources Agreement will reduce reliance on individual cases to adjudicate water use conflicts.

Great Lakes Restoration Initiative

A third major policy framework influencing water management in the Great Lakes region is the U.S.-based Great Lakes Restoration Initiative (GLRI). The GLRI is not a legal framework, but rather a U.S. federal policy initiative led by the U.S. Environmental Protection Agency in coordination with 10 other federal agencies. Beginning in 2010, the GLRI has become a formidable force for bringing in federal funds and leveraging, state, sub-regional, local and private funds to implement a suite of restoration priorities. A five-year GLRI Action Plan guides implementation: the current *GLRI Action Plan II* provides a roadmap for implementation from 2015 to 2019. The funding has been focused on four priorities: cleaning up legacy contaminated areas (known as Areas of Concern); preventing and controlling invasive species; reducing nutrient runoff that contributes to harmful/nuisance algal blooms; and restoring habitat to protect native species. While these priorities are not specific to water governance, GLRI implementation has had a significant impact mobilizing resources to improve water quality and more effectively manage water resources at the local/sub-regional scale.

---

[4] Article 504, Great Lakes Sustainable Water Resources Agreement of 2008.

Since its inception in 2010 and through FY 2016, GLRI has pumped over US$2.2 billion into the Great Lakes restoration. Many stakeholders on both sides of the Great Lakes would like to see an initiative like the GLRI on the Canadian side of the lakes; however, the political leadership and momentum have not been forthcoming.

Conclusion

Transboundary agreements in the Great Lakes basin reflect a mix of "hard" laws with enforcement and conflict resolution mechanisms and "soft" policies which have targets and deadlines but are not legally enforceable. A long history of amicable relations, multiple regional institutions, stable and similar national governments, and similar cultures provide the socio-psychological backdrop that supports transboundary governance. Increasingly common in the 21st century, formal regional institutions have developed cross-sectoral relationships and collaborations with tribes and First nations, environmental organizations, business associations and other elements of civil society.

Despite these strengths, threats to Great Lakes water resources are wholly different today than they were when the current suite of laws and policies governing Great Lakes water were established in the latter half of the 20th century. It is unclear whether the current legal and policy framework is sufficient to address emerging issues like climate change, industrial agriculture, and relationships between rural and urban communities who depend on the same water source for drinking, watering crops, and fishing. Transboundary cooperation is easier when there is plenty of water for everyone. Increased precipitation, storm intensity, water and air temperatures are already having and will continue to have cascading effects on hydrologic flow regimes, other ecological functions, and society at large. Whether and how riparian states and provinces adapt to altered flow timing and availability (e.g., floods and droughts) remains to be seen. Massive floods in southern Ontario and eastern New York in the 20-teens are becoming more common as 100-year floods are now occurring more frequently (Wendell 2017, Winkler et al. 2012). So far, these are being primarily addressed by the individual jurisdictions (states, provinces, municipalities). At the same time, the relative abundance of water across the region is also a challenge. Governments at all levels as well as civil society are less apt to advance water conservation, and its multiple ancillary benefits, when there is no compelling reason to do so. Finally, political leadership will continue to be a significant driver to effectuate the work of transboundary institutions.

References

Annin, P. (2006). Great Lakes Water Wars. Island Press, Washington, D.C., USA.

City of Waukesha Diversion Application. Secretariat of the Great Lakes-St. Lawrence River Water Resources Regional Body and Great Lakes-St. Lawrence River Basin Water Resources Council. Accessed online August 2017 at http://www.waukeshadiversion.org.

Cooley, H. and Gleick, P.H. (2011). Climate-proofing transboundary water agreements. *Hydrological Sciences Journal*, 56 (4).

Great Lakes Commission (2011). Annual Report of the Great Lakes Regional Water Use Database—Representing 2009 Water Use Data. Issue No. 18. Great Lakes Commission, Ann Arbor, Michigan, USA.

Great Lakes Regional Collaboration (GLRC). (2005). Strategy to Restore and Protect the Great Lakes. Accessed 10/19/12 from http://glrc.us/strategy.html.

Great Lakes Restoration Initiative (GLRI). (2010). Great Lakes Restoration Plan Initiative Action Plan, FY 2010-FY2014, and Great Lakes Restoration Plan Initiative Action Plan II, FY 2015-FY2019, (2014).

Hall, Noah, D. (2010). Interstate water compacts and climate change adaptation. *Environmental and Energy Policy Journal*, vol. 5.

Hildebrand, L.P., Pebbles, V. and Fraser, D.A. (2002). Cooperative ecosystem management across the Canada-U.S. border: approaches and experiences of transboundary programs in the Gulf of Maine, Great Lakes and Georgia Basin/Puget Sound. *Ocean and Coastal Management*, 45: 421–445.

Pebbles, V. (2014). Incorporating climate change adaptation into transboundary ecosystem management in the Great Lakes Basin. pp. 197–216. *In*: Sanchez, J.C. and Roberts, J. (eds.). Transboundary Water Governance: Adaptation to Climate Change. IUCN. Gland, Switzerland. ISBN 978-2-8317-1660-2.

Thoman, D., Pebbles, V. and Eddy, S. (2010). Great Lakes state and provincial climate change mitigation and adaptation: progress, challenges and opportunities. *Great Lakes Commission*, Ann Arbor, MI, USA.

Thurber, N.E., Water Level Management as an Option for Implementing the Coastal Zone Management Act in the Great Lakes Basin (n.d.), Great Lakes Coalition. Accessed 10/10/12 from http://www.iglc.org/water_management_czma.shtml.

US Army Corps of Engineers (USACE), Coordinating Committee on Great Lakes Basic. Hydraulic & Hydrologic Data (n.d.). Accessed 10/10/12 from http://www.lre.usace.army.mil/greatlakes/hh/links/ccglbhhd/.

US Environmental Protection Agency & Environment Canada (US EPA & EC). (1995). The Great Lakes: An Environmental Atlas and Resource Book. Accessed 10/10/12 from http://www.epa.gov/glnpo/atlas/glat-ch1.html#Understanding%20The%20Lakes.

Wendell, Joanna. (2017). Earth, Space and Science News. August 3, 2017. https://eos.org/articles/what-caused-the-ongoing-flooding-on-lake-ontario.

Winkler, J.A., Arritt, R.W. and Pryor, S.C. (2012). Climate projections for the midwest: availability, interpretation and synthesis. *In*: Winkler, J., J. Andresen, J. Hatfield, D. Bidwell and D. Brown (eds.). U.S. National Climate Assessment Midwest Technical Input Report. Available from the Great Lakes Integrated Sciences and Assessment (GLISA) Center, http://glisa.msu.edu/docs/NCA/MTIT_Future.pdf.

List of Laws/Agreements/Treaties

The Boundary Waters Treaty of 1909. Treaty between the United States and Great Britain relating to boundary waters, and questions arising between the United States and Canada. Accessed 10/19/12 from http://bwt.ijc.org/index.php?page=Treaty-Text&hl=eng.

Great Lakes Compact of 1955. Created through the collective legislative action of the Great Lakes states and later granted congressional consent through United States Public Law 90-419 established the Great Lakes Commission to promote the orderly, integrated, and comprehensive development, use, and conservation of the water resources of the Great Lakes Basin". Accessed 11/21/12 from http://www.glc.org/about/glbc.html.

Great Lakes Restoration Initiative (GLRI). (2010). A United States-based policy initiative managed by a task force of 11 federal agencies that coordinate to develop and implement the GLRI Action Plan covering five priority "focus areas". Accessed from http://greatlakesrestoration.us/index.html.

Great Lakes-St. Lawrence River Basin Water Resources Compact of 2005. U.S. Public Law 110–342—October. 3, 2008. http://www.gpo.gov/fdsys/pkg/PLAW-110publ342/pdf/PLAW-110publ342.pdf.

Great Lakes-St. Lawrence River Basin Sustainable Water Resources Agreement and Compact of 2005. Signed by the Great Lakes Governors and Premiers at the Council of Great Lakes Governors on December 13, 2005. Accessed 10/10/12 from http://www.cglg.org/projects/water/CompactImplementation.asp.

Great Lakes Water Quality Agreement of 1978 (amended in 1983, 1987 and September 2012). Agreement between the United States of America and Canada. Accessed 10/10/12 from http://www.epa.gov/grtlakes/glwqa/.

United Nations Treaty Collection at http://treaties.un.org/Pages/ViewDetails.aspx?src=UNTSONLINE&tabid=2&mtdsg_no=XXVII-12&chapter=27&lang=en#Participants. Accessed October 19, 2012.

United States Executive Order 13340 of May 18, 2004. Created the Great Lakes Interagency Task Force and directed the U.S. Environmental Protection Agency (EPA) to convene a regional collaborative, known as the Great Lakes Regional Collaboration.

CHAPTER 5

Legal Aspects of Transboundary Lakes Governance in the Western Balkans

Slavko Bogdanovic

Introduction

This chapter has been designed to comprise current legal aspects of governance systems applied or under development in the catchment areas (not withstanding the specific legal status of the areas declared) of several natural freshwater lakes in the southern part of the Balkan Peninsula, shared by littoral states. These are Skadar/Shkodra (shared by Albania and Montenegro), Ohrid (shared by Albania and Macedonia), Macro and Micro Prespa (shared by Albania, Greece and Macedonia) and Dojran (shared by Macedonia and Greece) lakes. All these lakes, except the Dojran Lake, are part of the Drin River basin, which, besides these four lake sub-basins comprises also the geographical areas (sub-basins) of the Black Drin River (*Crn Drim/Drin i Zi*), the White Drin River (*Beli Drim/ Drin i Bardhë*), the Drin River (*Drim/Drini/Drin i madh*) and the Bojana/ Buna River. The established denomination for the entire geographical area is "the Extended Transboundary Drin Basin", is shown in Map 5.1.

Some legal aspects of the notion "governance" shall be described briefly and followed thereafter with a concise review of the most important legal instruments applicable at the level of the lake basins in review as well as a description of the governance concepts identified. In

---

Ilije Ognjanovica 4, 21000 Novi Sad, Serbia.
Email: nsslavko@gmail.com

Legal Aspects of Transboundary Lakes Governance in the Western Balkans 79

Map 5.1. The Extended Transboundary Drin Basin.

concluding remarks, those legal/institutional elements of the systems of lake governance that could be identified as features of a desirable "good" transboundary waters governance practice will be highlighted.

The Concept of Governance

In a modern society, the role of state in natural resources/environment preservation, improvement and protection, including water administration/management, could be defined as the role of a "trustee for its citizens", steward or guardian. This role is naturally rooted in the (territorial) state sovereignty over natural resources, regardless of whether they are defined (declared) to be in a state ownership or in a type of the public domain or public ownership. However, this role of the state does not necessarily mean the application of its *ius imperium* in management of those assets (like in other sectors of administration, e.g., security/defence, interior affairs, etc.). As a trustee, steward or guardian the responsible state authorities (state administration) are subject to public control (in terms of the notion "public-right-to-know"), which implies the wide involvement of the public and local autonomous bodies in the decision making process.[1] The Aarhus Convention, a unique multilateral international treaty that connects environmental protection and human rights principles,[2] is a worldwide paradigm on core elements of environmental governance that also comprise transboundary waters. This seems to be a broadest blueprint for elaboration of various approaches to such notions as management, governance and good governance of natural (and/or water) resources.[3]

The good governance concept may be seen as a management/administrative concept, which comprises participation of the state's authorities (on all institutional levels) and civil society (and all its organizations) and relations between them. The concept is closely connected to the concept and practice of democracy and to the way a government use willpower of the people to ensure wellbeing of all citizens.[4]

---

[1] Bogdanovic (1993).
[2] AARHUS CONVENTION (1998).
[3] For more details on the subject see Mici, A. (2016), Web (22.12.2016). Notwithstanding the fact that this chapter deals exclusively with the legal aspects of transboundary lakes management, it is worthwhile to draw attention to this presentation, which is much broader than a pure lawyer's approach, as a concise review of elements important for identification of contents and features of certain concepts (governance, management, environmental governance – including IUCN approach, good governance), differentiating possible types of governance and concepts of governance and management, specifically pointing out the importance of public involvement, etc.
[4] Fernandez-Crespo (2008).

In the times before the fall of Berlin Wall (1989), the Balkan socialistic states practised a rigid concept of state ownership over all their natural resources, without adequate inclusion of stakeholders (public, interested public, local communities) in the process of water management. Perhaps, only the Socialistic Federal Republic of Yugoslavia – SFRY has had a "softer" approach in this sense, stemming from its federal structure and self-management social system in place there. Nevertheless, rigidity in this field quickly appears in all new states after recomposition of the SFRY territory into new independent states (at the beginning of the 1990s). The state authorities appeared as key and the only players in the field of water management, acting in the name of the state, or in the name of people and their citizens.[5]

Drivers of Change

Such rigid perception and practice of the role of the state in managing of national (and consequentially shared) water resources started to change under the strong influence of a new, "shifted" paradigm, applicable to transboundary water resources in the region of Europe (and in a broader context than Europe, today), developed in several multilateral treaties (conventions) under the aegis of UNECE during the 1990s.[6] Speaking about water only, fundamentally established, but still open for development and improvements, by these (mutually harmonised) treaties, these international

---

[5] Bogdanovic (1993), *ibid.*
[6] The most important in this context are: the Convention on the Protection and Use of Transboundary Watercourses and International Lakes – UNECE Water Convention (Helsinki 1992); the Convention on Environmental Impact Assessment in a Transboundary Context – EIA, the Espoo Convention; Protocol on Strategic Environmental Assessment to the EIA Convention (Kiev 2003); The Convention on Access to Information, Public Participation in Decision-Making and Access to Justice in Environmental Matters – the Aarhus Convention (Aarhus 1998). This huge policy and legislative move in Europe followed a common acceptance of the Parties to the Conference on Security and Co-operation in Europe – CSCE (later Organisation for Security and Co-operation – OSCE), in its Final Act (Helsinki 1975), of the existence, in international law, of a duty to prevent activities on-going in the territory of one member state to be cause of environmental degradation in the territory of another member state (Birni and Boyle 2002). Regarding water pollution control and fresh water utilization, the Parties to the Helsinki Final Act specifically accepted, *inter alia*, that the aims of their co-operation shall be studying, increasing effectiveness, taking necessary measures and encouragement of national and international efforts for prevention and control of water pollution, in particular of transboundary rivers and international lakes, development of techniques for the improvement of the quality of water and further development of ways and means for industrial and municipal sewage effluent purification. Inclusion of the issue of co-operation between the states sharing the same (natural) water resources, in a context of international security, induced a new approach to management of water resources. By requiring due care of international security aspect of the utilization of shared resources, that approach led to development of a multi-layer governance system in Europe, established by an (eventually structured, in terms of its coherency and internal consistency) number of policy instruments and legal treaties in the region of Europe.

law frameworks are naturally oriented toward bringing changes in management patterns, introducing an evolving, "shifted management paradigm".[7] By their territorial and functional scope, or by their nature of international law, this system of treaties span from covering the entire UNECE region, watersheds (Baltic and Black sees, or Mediterranean see, including Adriatic), or certain river/lake catchment area. They can be multilateral, bilateral, trilateral and can comprise only (exclusively) co/operation of the competent authorities of the involved states.However, there are treaties that include in such co/operation (even in decision-making) a remarkably broad range of various categories of stakeholders.[8] This body of international law has become one of two most powerful drivers of changes in the management of transboundary waters in contemporary Europe.

Another strong driver of change of the water management patterns, fully in harmony with the entire UNECE legal context, is the legal system of the European Union (EU) applicable on water resources (national and shared), which should be incorporated into the national legal systems. Except Greece, other western Balkan countries under the review in this chapter (littoral countries) are on their way to join EU.[9] Therefore, they are committed to approximation to EU water *acquis*, i.e., to gradual transposition, implementation and enforcement, among other environmental subsectors, the EU policy and legislation regarding waters,[10] and nature,[11] with the aim of full compliance before the date of becoming the EU member states. The approximation of national

---

[7] Which, to mention some core features, comprise of integrated management of water resources (including all uses of waters and taking into account all users' interests) in a river/lake basin area, and having in view all other water management aspects (protection against detrimental effects from surface and underground waters, protection of quality of waters in water bodies, preservation and protection of water ecosystems and ecosystems dependent of waters, etc.).

[8] Local self-government units, NGOs, scientific organizations, local communities/groups of citizens, INGOs and other international community actors interested, etc.

[9] They have the status of countries candidate for the membership in EU, on the basis of Stabilisation and Association Agreements (SAA) signed with EU. Macedonia has gained the status of a candidate country on 16.12.2005, Montenegro on 17.12.2010, and Albania on 27.06.2014.

[10] The EU water *acquis* consists of a number of legislative acts (directives and regulations), regulating numerous aspects of water quality protection, including very detailed and demanding monitoring (for all inland surface freshwater and groundwater bodies, coastal and transitional waters), including for bathing waters (Directive 2006/7/EC) and urban waste waters (UWWT Directive 91/271/EEC). River/lake basin has been established as the managerial unit, which a complex management plan has to be developed and adopted for (WFD-2000/60/EC). Water quality (environmental) objectives and standards are set-out (Directive 2008/105/EC), as well as obligation for development of various action plans aimed at their application and achievement. Additionally, assessment and management of flood risks is also regulated (Directive 2007/60/EC), etc.

[11] Mentioned here should be made of the Directive 92/43/EEC (habitat), Regulation (EC) No. 338/97 (wild flora and fauna) and Directive 2009/147/EC (wild birds).

legislation is a binding international duty of the countries candidate for the membership in EU, established by an international law instrument – SAA. Fulfilling this duty is subject to national strategic planning, in the form of a governmental approximation strategy, and achieving the full compliance with the EU *acquis*, through implementation of an action plan containing the measures to be undertaken, designation of responsible authorities/bodies and time frames, as well as other conditions (e.g., need for international aid) for application of those measures.[12]

There are several global multilateral legal instruments relevant here, of which the most important, for the region in review, is the Ramsar Convention.[13] Under the aegis of the Med-Wet Initiative that includes participation of some 20 countries, this global environmental treaty is particularly important for co-operation in the Prespa Lakes region.[14]

Relations between the countries concerning the shared (catchment areas of) lakes in review, historically speaking, from the legal perspective, were really limited. In the times of Socialistic Federal Republic of Yugoslavia (SFRY), i.e., roughly speaking until the fall of Berlin Wall (1989), there were only a few specific bilateral treaties.[15] From the standpoint of SFRY, those treaties can be seen as a part of a broad political effort of SFRY as a non-aligned country in the divided bipolar world, aimed at establishing a long-term co-operation with its neighbours on various issues related to transboundary waters, in the framework of its foreign policy based on the principles of peaceful co-existence and friendship among peoples of neighbouring states, and permanent attempts of SFRY to provide conditions for development and benefit of all its (including

---

[12] An example of such a strategy, is the Montenegrin National Strategy for Transposition, Implementation and Enforcement of EU *acquis* on Environmental and Climate Change (with an Action Plan) for the period 2016–2020. GOVERNMENT OF MONTENEGRO (2016).

[13] UN Convention on Wetlands of International Importance, especially as waterfowl Habitat (Ramsar 1971).

[14] For more details, see in Bogdanovic (2008).

[15] Those were: Agreement between the Government of the Federal People's Republic of Yugoslavia and the Government of the People's Republic of Albania on Water Economy Issues (1956), pertaining to the obligation of the Parties to consider and jointly resolve all the issues regarding water economy, measures and works on shared watercourses, lakes and hydraulic systems, particularly to the Ohrid Lake, Back Drin, White Drin, Skadar Lake and the Bojana River (the last signs of its application had been see in 1988); Agreement between the Federal People's Republic of Yugoslavia and Kingdom of Greece on Water Economy Issues, which had comprised the co-operation between two states particularly regarding studies of the Vardar/Axios River, as well as issues (including water balance, torments in the bordering area, fishery in two lakes, etc.) regarding the Dojran and Prespa lakes (which was terminated by exchange of the notes 10.12.1997); Agreement between the SFRY Government and the Government of the Kingdom of Greece connected with development of the study on integral bonification on the catchment area of the Vardar/Axios River, which was aimed at obtaining funding by UNDP (which was terminated by exchange of the notes 10.12.1997).

84 *Lake Governance*

bordering) regions.[16] The history of bilateral/neighbouring relations regarding the shared lakes in this region, lasting as long as 60 years shows no remarkable results in founding those relations on mutual confidence of nations involved and on lasting and sustainable legal grounds.

Yet, on-going multi-year activities in this field, in the post-fall of the Berlin Wall era, with the permanent presence of the international community,[17] with strongly defined policy elements[18] and a robust financial aid, with an enabling environment in which sustainability of governance/management of these pristine WB/European water (natural) assets seem to be a reachable objective. Particularly if a broader political context, which, in certain aspects, has been generating (so far) an environment impeding the EU and NATO integration prospects of Macedonia – begins to melt down.

Core elements that make the legal and policy matrix applicable on the lake governance of WB are, presented in Table 5.1.

---

[16] More on the water treaties see in Bogdanovic (2005).
[17] Such as: EU, WB, GEF, UNDP, German, Swiss and Greece governments, Med-Wet Initiative.
[18] Besides EU policy in the field of water, the most important and influential are specific policy approaches, developed by German and Greece Governments in the framework of Petersberg Phase II/Athens Declaration Process, a regional Dialogue for Transboundary water management in South Eastern Europe (TWRM 2012). Web (25.12.2016): http://www.twrm-med.net/southeastern-europe. The approach assuming the "use of water as a catalyst for regional co-operation rather than a source of potential conflict", was designed by the Petersberg Declaration, which is the result of the International Dialogue Forum "Global Water Politic—Co-operation for Transboundary Water Management" (Petersberg, Germany; 03–05 March 1998). The Forum was a collaborative effort of the German Federal Ministry for Economic Co-operation and Development (BMZ), the Federal Foreign Office (AA), The Federal Ministry for the Environment, Nature Conservation and Nuclear Safety (BMU), the World Bank and Development Policy Forum (EF)/German Foundation for International Development (DSE) (TWRM 2012). Web (25.12.2016): http://www.twrm-med.net/southeastern-europe/regional-dialogue/framework/petersberg-phase-ii-athens-declaration-process/petersberg-process/Petersberg%20declaration.pdf. Athens Declaration "Actions to Promote Sustainable Management of Transboundary Water Resources in the South-Eastern Europe and Mediterranean Region", resulted from the International Conference on "Sustainable Development for Lasting Peace: Shared Water, Shared Future, Shared Knowledge", organized by the Hellenic Presidency of EU, Hellenic Ministry of Foreign Affairs and Hellenic Ministry for Environment, Physical Planning and Public Works and World Bank (Athens 2003). The Declarations calls, *inter alia*, for a greater emphasis, in management of shared waters, on making their use sustainable through involving citizens and CSO. It also calls for negotiations and reaching agreements between co-operating parties through legally binding instruments. However, according to the Declaration, such legally binding instruments (international treaties) should follow after, while "a wide range of planning and management activities can [...] be initiated prior." (TWRM 2012). Web (25.12.2016): http://www.twrm-med.net/southeastern-europe/regional-dialogue/framework/petersberg-phase-ii-athens-declaration-process/athens-declaration/English.pdf. This approach can be seen as implying continuation of the lack of political will (as an obstacle) for concluding binding instrument(s) of international law, and, as a consequence, not ensuring sustainable funding (from budgetary sources of littoral countries) of transboundary related co-operation.

Table 5.1. Applicable Legal & Policy Instruments (International & National).

Instrument Title	Signatories & Parties					Notes
	AL	GR	MAC	MN	EU	
Global treaties (UN system)						
UN Convention on Wetlands of International Importance, especially as waterfowl Habitat (Ramsar 1971)	1994 P	1975 P	1991 P	1977 P	n/a	
UNESCO Convention Concerning the Protection of the World Cultural and Natural Heritage (Paris 1972)	1989 P	1981 P	1974 P	1974 P	n/a	
UN Convention on International Trade in Endangered Species of Wild Flora and Fauna (CITES; WASHINGTON D.C. 1973)	2003 P	1992 P	1999 P	2001 P	n/a	
UN Convention on Biological Diversity (Rio de Janeiro 1992)	1994 P	1994 P	1994 P	2001 P	1993 P	
UN Convention on the Law on Non-Navigational Uses of International Watercourses (New York 1997)	1997 S	1997 S	1991 S	2013 P	n/a	
Council of Europe (CoE) treaty						
Convention on the Conservation of European Wildlife and Natural Habitats (Bern 1979)	1998 P	1993 P	1999 P	2008 P	1982 P	
UNECE treaties						
Convention on Environmental Impact Assessment in a Transboundary Context (Espoo Convention; Espoo 1991)	1991 P	1998 P	1999 P	2009 P	1997 P	
Convention on the Protection and Use of Transboundary Watercourses and International Lakes (Water Convention, Helsinki 1992)	1994 P	1996 P	2015 P	2012 P	1996 P	

Table 5.1 contd. ...

...Table 5.1 contd.

Instrument Title	Signatories & Parties					Notes
	AL	GR	MAC	MN	EU	
UNECE treaties						
Convention on Access to Information, Public Participation in Decision-Making and Access to Justice in Environmental Matters (Aarhus Convention, Aarhus 1998)	2001 P	2006 P	1999 P	2003 P	2001 P	
Protocol on Strategic Environmental Assessment (SEA, Kiev 1993)	2005 P	2003 P	2003 P	2009 P	n/a	
Convention for Protection of the Mediterranean Sea against Pollution (Barcelona Convention 1976)	1990 P	n/a	n/a	2007 P	1978 P	
Western Balkans lake treaties						
Agreement between the Council of Ministers of the Republic of Albania and the Government of the Republic of Macedonia for the Protection and Sustainable Development of Lake Ohrid and its Watershed (Skopje 2004)	S	n/a	S	n/a	n/a	
Agreement on the Protection and Sustainable Development of the Prespa Park Area (Pily 2010)	P	S	P	P	P	
Western Balkans "soft law" & transboundary governance related instruments						
Memorandum of Understanding for Cooperation in the Field of Environmental Protection and Sustainable Development between Albania and Montenegro (2003)	S	n/a	n/a	S	n/a	
Memorandum of Understanding (MoU) of 29 November 1996 between the Albanian and Macedonian Governments regarding the Lake Ohrid Conservation Project						Not available

Table 5.1 contd. ...

...Table 5.1 contd.

Instrument Title	Signatories & Parties					Notes
	AL	GR	MAC	MN	EU	
Western Balkans "soft law" & transboundary governance related instruments						
Memorandum of Understanding of 7 September 2000 concerning Cooperation in the Field of Environmental Protection and Sustainable Development between the Macedonian Ministry of Environment and Physical Planning and the Albanian National Environmental Agency						Not available
Declaration on the Drin Strategic Shared Vision (October 2011)						Kosovo is also the Signatory
Memorandum of Understanding for the Management of the Extended Transboundary Drin Basin – THE DRIN STRATEGIC SHARED VISION (25.11.011)						Kosovo is also the Signatory
Petersberg Declaration (1998)	-	-	-	-	-	Germani & WB
Athens Declaration (2003)	-	-	-	-	-	EU, Greek, & WB
Declaration on the Skadar/Shkodra Lake (01.02.2006)	S	-	-	S	-	Various stakeholders
Declaration on the Creation of the Prespa Park and the Environmental Protection and Sustainable Development of the Prespa Lakes and their Surroundings (Agios Germanos, 02.02.2000)	S	S	S	-	-	
Joint Statement [of the Prime Ministers of AL, GR, Mac] on the Prespa Lakes Basin (Pily, 27.11.2009)	S	S	S	-	-	

S – Signatory;
P – Party;
n/a – not applicable

National Legal Context

Since the middle of the 20th century, the management (or governance) of the lakes and their surroundings under review in this chapter, has been subject of a specific attention of the littoral states, from more different aspects. Awareness of the exceptional natural values in this area led to designation of first national parks in Yugoslavia, through the forestry legislation. Management of the waters in the lakes catchment areas has been generally subjected to water laws in all littoral countries. However, a special law regarding protection of the lakes in Macedonia was enacted in 1977.[19] Albania also enacted a law specifically regulating only protection of its transboundary lakes.[20] National legislation on environmental protection (preservation and improvement), as well as on forestry, in all littoral countries, which has been developing continually in last 30 years, under the influence of drivers indicated below, have also played a role in the governance of these lakes.

All respective national systems of legislation, regulating water, nature and environmental management/governance issues in Albania, Macedonia and Montenegro are passing their specific transition from a socialistic legacy and attempts to satisfy the current requirements of a new democracy and market paradigms towards compliance with (and playing role in) an evolving, very demanding, multi-layer environmental (including water) governance. It is a matter of fact that a full awareness of the global, UNECE, EU and subregional (in terms of bilateral, trilateral, river/lake basin, etc.) aspects of the lake areas governance is present, along with numerous attempts to make moves forward. Also, it is clear that respective legislation needs to be further streamlined with the EU paradigm, in terms of full transposition of the EU requirements, implementation of harmonized legislation and its enforcement. The international community[21] has been present in the region permanently for years providing support and guidance in developing, a sustainable necessary governance system, scientific information, strategic planning and policy grounds, and working on building clear and long lasting commitments of key actors. Commitments that also include acceptance of binding treaties.

It should be noted that, in doing their "homework" in the transboundary areas which their lake catchment areas covers, littoral countries are facing serious challenges, that can be attributed to the political legacy, limited mutual confidence, economic underdevelopment

---

[19] Law on protection of the Ohrid, Prespa and Dojran lakes ("Official Gazette of SRM", No. 45/1977).
[20] Law No. 9103 of 10.07.2003 on the protection of transboundary lakes.
[21] *See supra*, note 17.

and poverty, etc. Those challenges comprise the need for tackling with a wide range of issues that include, e.g., infrastructure development without considering sustainable use of natural resources, unharmonized fishing policy, unsustainable forestry and hunting, unsustainable tourism, gravel extraction/as a matter of concern, lack of implementation of plans developed with international aid support, continuing pollution problems, need for harmonized monitoring programmes.[22] Even in protected areas uncontrolled building activities are noticeable, as well as illegal hunting and logging, which indicates limited power or lack of law enforcement (due to the limited financial means of competent authorities, weak institutional setting, etc.).[23] Generally, in all littoral countries, an insufficient management of sub-basin management both at national and transboundary level was indicated.[24] In lieu of permanent joint efforts of the international community, national administrations, local stakeholders and NGOs, it seems that desirable change is hardly visible.

Specific Governance Regimes

1. Ohrid Lake Agreement

The Ohrid Lake has been considered as one of the 17 ancient world lakes, and the oldest one in Europe. Its formation followed tectonic activity(ies) some 4–10 million years ago. The lake is shared between Albania (1/3) and Macedonia (2/3). In its catchment area there are 40 tributaries (creeks and rivers, that flow temporarily, depending on precipitation). This hydrological system (sub-basin) belongs to the Drin River watershed, which empties into the Adriatic Sea. There is a natural underground connection between Prespa (Macro) Lake and the Ohrid Lake, which enables flow of the Prespa Lake waters into the Ohrid Lake.[25] The geographical position of the Ohrid Lake can be seen in Map 5.2—Ohrid Lake.

Due to its exceptional natural and cultural values, the Ohrid Lake has been declared as a UNESCO World Heritage site – natural (since 1979) and cultural (since 1980) and UNESCO transboundary biosphere reserve since 2014. There are on-going efforts by international and local players to extend this status to the Albanian part of the lake, and to reinforce transboundary co-operation between the two littoral countries sharing the lake.[26]

---

[22] Skarbøvik et al. (2008).
[23] ROYAL HASKONING (2006).
[24] Skarbøvik et al. (2008).
[25] For more details *see*: Web (22.12.2016): http://www.worldlakes.org/lakedetails.asp?lakeid=8770.
[26] IUCN (2014), 07.10.2014.

90 *Lake Governance*

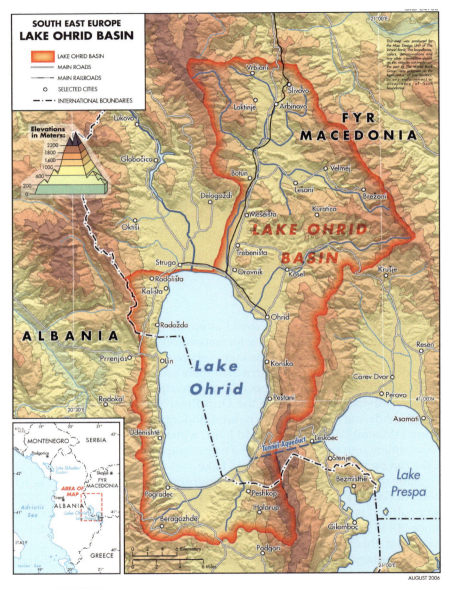

Map 5.2. Ohrid Lake.

Track records on relations between Albania and Macedonia regarding the Ohrid Lake shows that their co-operation have been initiated and is gradually developing and setting up three legal instruments, signed in the last 20 years.[27] With a clear awareness of local, regional and global significance of the Ohrid Lake as a site of exceptional environmental, economic, scientific, aesthetic and cultural value[28] the two Governments committed themselves to work on achieving three basic objectives:

- Assuring integrated protection and sustainable development of the Lake and its watershed, in accordance with EU standards;
- Getting the World Cultural and Heritage status and the UNESCO status of "Biosphere Reserve"; and
- Providing conditions for designation of the Prespa-Ohrid region as one of the only two Euro-regions in SEE.[29]

For the purpose of the OA, "the Ohrid Lake watershed" was defined as comprising surrounding catchment areas, including Prespa (Macro and Micro) Lakes and their respective watersheds in the territories of two parties. Several joint bodies are identified by OA[30] which have been entrusted by the Parties with land-use or environmental management functions or responsibilities in the Ohrid Lake Watershed.[31]

In order to fulfil the Ohrid Agreement objectives, the Parties accepted to undertake necessary measures aimed at prevention, control and reducing of pollution of the waters and protection of soil from erosion, depletion, infections and pollution in the watershed, protection of biodiversity, prevention of introduction of non-autochtonous animal and plant species, ensuring sustainable use of natural resources, avoiding serious damage to cultural heritage and landscapes, and preventing and controlling economic activities which have a negative impact in the watershed. With the view of preventing and controlling impact of economic activities, the OA Parties agreed to harmonize their criteria and standards, strategies and regulations

---

[27] Firstly, the Memorandum of Understanding (MoU) of 29 November 1996 between the Albanian and Macedonian Governments regarding the Lake Ohrid Conservation Project, was signed, and then the Memorandum of Understanding of 7 September 2000 concerning Co-operation in the Field of Environmental Protection and Sustainable Development between the Macedonian Ministry of Environment and Physical Planning and the Albanian National Environmental Agency. Finally, co-operation between two littoral states of the Ohrid Lake, based on these two MoUs, led to signature of the Agreement between the Council of Ministers of the Republic of Albania and the Government of the Republic of Macedonia for the Protection and Sustainable Development of Lake Ohrid and its Watershed, in Skopje, on 17 June 2004 (Henceforth: Ohrid Agreement and OA).
[28] Ohrid Agreement (2004), Preamble.
[29] Id.
[30] Those are: Established by the Ohrid Lake Conservation Project, which include the Ohrid Lake Monitoring Task Force, the Watershed Management Committee and the Organization of the Fishery Management, as well as the Prespa Park Coordinating Committee and any other body designated by the Parties.
[31] Ohrid Agreement (2004), Article 2.

pertaining to spatial and urban plans and land management plans, ensure full application of individual rights regarding environmental issues, set out by the Aarhus Convention and establish and maintain an effective monitoring system in order to keep under the control environmental state and quality of the Lake and its watershed.[32]

The Parties to the Agreement agreed to establish and faithfully apply environmental standards, exact criteria, limits and objectives for the protection, conservation and sustainable development of the Ohrid Lake and its watershed, with assistance of the Ohrid Lake Watershed Committee established by the Agreement. For that purpose, they accepted to "establish separate technical documents that (a) implement the legal commitment toward the application of international norms and standards and the ones of the European Union" for the protection and conservation of the Lake and its watershed, adjust relevant national environmental standards and criteria to local conditions, and require use of "best available technology and contemporary practices".[33]

With the aim of achieving its objectives, a bilateral institution, the Ohrid Lake Watershed Committee (OLWC), was established by OA, which is composed of equal number of members from each Party. Determination of the members of the OLWC includes "three titulars of central governmental institutions", appointed by respective Governments, "three titulars of local governments institutions" and one representative of the civil society. There is also one non-voting representative of the international donor community. There are no more detailed provisions indicating who has authority to appoint the "titulars" from local governments and civil society (which part of civil society) and how the international donor community would appoint its representative. Experts might be invited on the OA meetings too.[34] The functions of OA are exhaustively listed and comprise monitoring of activities carried out for the protection of the lake and its watershed, submitting suggestions for undertaking of necessary measures and activities for implementation of OA to the OA Parties and inviting them to co-operate, to co-ordinate and carry out joint projects, make evidence of actions and attitudes of the OA Parties in contradiction with OA.[35] The specific functions of OLWC have been formulated as follows:

- Drafting and application of standards, environmental criteria and the requirements of sustainable development;
- Completing the legal regulatory framework of the watershed area;

---

[32] Ohrid Agreement (2004), Article 3.
[33] *Op. cit.*, Article 4.
[34] *Op. cit.*, Article 5.
[35] *Op. cit.*, Article 5.

- Drafting and application of strategies, programmes and action plans to be implemented;
- Drafting the programme and the application of effective systems of monitoring in order to keep under control the state of environment and the quality of the lake and its watershed;
- Gathering, elaboration and publication of relevant environmental information;
- Preparation of the activities for creating conditions for designation of the Prespa—Ohrid watershed as one of only two Euro-regions in South Eastern Europe (SEE);
- Increasing of public, NGO-s and other stakeholders' participation.[36]

OLWC submits recommendation and opinions to the Parties, based on previously provided of (mutual) bilateral sub-commissions in specific areas, regarding relevant spatial plans, "status of legislative and regulative" measures and relevant programmes for development, policy and decisions of the governmental and local self-government authorities.[37]

The regular meetings of OLWC should be scheduled at least twice a year (once every six months), while an extraordinary meeting should be called on the request of a simple majority of its members. The meetings are to be held interchangeable in Albania and Macedonia. The OLWC Chairman and the Secretary are to be appointed at the first OLWC meeting. Their term of office lasts one year and the Parties representatives at those functions rotate each year. The OLWC scope of competence has been strongly determined as serving "in the capacity of an intergovernmental body, which keeps relations with donors in order to gain projects and donations" to be used for the OA implementation. OLWC has an obligation to prepare and publish an annual report on the state of environment in the watershed and its own work.[38] OLWC shall take its decisions on recommendations and opinions by consensus. If the consensus has not been reached a non-settled issue shall be submitted to the attention of two governments (Article 7 para 1). Its decisions (on recommendations and suggestions), by their definition, shall be addressed to the Parties (in the form of) "recommendations by setting time limits" for their implementation by the Parties. The OA Parties accepted obligation to implement the OLWC decisions in accordance with their national law, and to report back regularly to the OLWC on the measures taken (or not taken, including details on reasons and proposal for way and time of the implementation).[39]

---

[36] *Op. cit.*, Article 6.
[37] *Op. cit.*, Article 6.
[38] *Op. cit.*, Article 5.
[39] *Op. cit.*, Article 7.

OLWC has its Secretariat, located in the City of Ohrid, Macedonia, composed of one member of each Party at least, the head of which is the OLWC Secretary. The Secretariat has a status of a technical body dependent on OLWC and acting in the name of OLWC. Its functions comprise *inter alia* preparation of the OLWC meetings, drafting administrative documents and decisions, gathering and processing data and information, preparing studies, analyses and projects, publish and disseminate materials and keep relations with all stakeholders. The Secretariat has the duty to submit its reports to the each OLWC meeting (it means twice a year). The Macedonian Party accepted to provide premises, equipment and all necessary technical goods and means needed for the work of the Secretariat, while each Party bears financial costs for work of their representatives, etc.[40]

OA also contains short provisions regarding its amending, dispute settlement (by means of negotiation or any other means acceptable to them or through diplomatic channels), its entry into force, duration and termination and authentic texts.

2. *The Prespa Park*

Prespa is also one of the ancient lakes, estimated to be more than five million years old.[41] There are two Prespa lakes, which are the highest tectonic lakes in the Balkans (at 852 m a.s.l.). The Great Prespa Lake, with surface area 259 km^2 is been shared between Macedonia (176.3 km^2), Albania (46.3 km^2) and Greece (36.4 km^2). The Small Prespa Lake is shared between Greece (42.5 km^2) and Albania (4.3 km^2).[42] These two lakes are connected. There is an underground connection in the carstic soil between the Great Prespa Lake and Ohrid Lake, making in that way the Prespa Lakes a part of the Drin River basin catchment area.

As a consequence of historical realities (to mention here the Greek Civil War, which ended in 1949 and the totalitarism of Enver Hoxa in Albania) the area along the lakes shores was depopulated, and under the status of "no-man's land" in Albania (1949–1990), and a special military area, in Greece (1949–1973) with controlled movement (in and out) of population, a long time after the Civil War. It was only in the 1970s, when tourism started to be promoted there.[43] The Prespa Lakes region encompass particularly poor bordering areas in Albania and Macedonia limiting possibilities for co-operation, while the Greek side of the picture

---

[40] *Op. cit.*, Article 8.
[41] Web (22.01.2017): http://www.worldlakes.org/lakedetails.asp?lakeid=9176.
[42] Web (22.01.2017): https://en.wikipedia.org/wiki/Lake_Prespa.
[43] Gardin, J. (2015); LAKEPEDIA (2015); Web (22.01.2017): http://www.lakepedia.com/lake/prespa.html.

includes a hard longstanding strategy – beyond the field of environmental protection, no more co-operation is wanted with neighbours.[44]

According to certain authors, the specific biodiversity of Prespa Park is important so is the biodiversity of Germany as a whole.[45] In Macedonia, National Parks Pelister and Galicica were declared in 1948 and 1958, respectively. In Greece, the Prespa National Park was declared in 1974, while in Albania, the National Park was declared in 1999. Parts of the Prespa area in Greece and Macedonia were established as Ramsar sites 1974 and 1995 respectively. In Macedonia, the Ezerani Strictly Protected Reserve was established since 1996, while the Great Prespa Lake was designated as a National Monument of Nature (IUCN category III) since 2011. In 2014 the UNESCO Transboundary Biosphere Reserve was established, in the territories of Albania and Macedonia.[46] Visual imagery of the Prespa Lakes Basin is presented in Map 5.3.

Recognizing that Prespa Lakes and their surrounding catchment area have a significant international importance due to their geomorphology, ecological wealth and biodiversity, which provide habitat for various and rare species of flora and fauna, offer refuge for migratory bird populations, and constitute a much needed nesting place for many species of birds threatened with extinction, the Prime Ministers of Albania, Greece and Macedonia signed the Prespa Park Declaration.[47] By signing the Declaration, the PMs also recognized that conservation and protection of the ecosystem of such importance, not only renders a service to Nature, but also creates opportunities for the economic development of adjacent areas in the littoral countries and proves compatibility of traditional activities and knowledge with conservation of nature.[48] The Declaration contains also their (rather weak) commitment, expressed with words "joint action would be considered [...]", which, nevertheless, led to the remarkable achievements in co-operation in the following decade,[49] including creation of the Prespa Park Co-ordination Committee (PPCC) and one more declaration of the three prime ministers[50] that eventually led to the signing of a binding treaty – the Prespa Park Agreement.

---

[44] Gardin, J. (2015).
[45] *Op. cit.*, Having in view the complex political, economic and environmental (ecological) aspects of the relations of littoral states in this region, and strongly pointing out the important role of the Greek NGO Society for the Protection of Prespa (SPP). (http://www.spp.gr/index.php?option=com_content&view=article&id=10&Itemid=15&lang=en), this author indicates that "the language of the environment will not translate the difficult issues of the political, cultural and economic situation into pacified objects of negotiation."
[46] SSP (2017); Tasevska, O. et al. (2016).
[47] Prespa Park Declaration (2000), first paragraph.
[48] *Op. cit.*, Second paragraph.
[49] Bogdanovic (2008).
[50] PMs Joint Statement (2009).

96 *Lake Governance*

Map 5.3. Prespa Lakes.

The Agreement on the Protection and Sustainable Development of the Prespa Park Area has been signed[51] by the countries littoral to the Macro and Micro Prespa Lakes, Albania, Greece and Macedonia and European Union, as the fourth party. EU joined this agreement due to the fact that the Agreement covers matters that are in the EU jurisdiction, that one of the parties is the EU member state (Greece), while one is the candidate country for the EU membership (Macedonia) and one another (Albania) is a potential candidate.[52] The Agreement covers the Prespa Park Area, which has been defined as the geographical area in the territories of three states, that share the basin of the Prespa Lakes. The Prespa Park Area comprises the surface and "properly assigned groundwaters", and it is designated as a transboundary protected area, under the Declaration of 2 February 2000.[53,54] The objective of the Agreement is ensuring integrated protection of the ecosystem and sustainable development of the Prespa Park Area, through the co-operation of the Parties.[55]

Besides a clear formulation of the basic obligation of the Parties[56] the Agreement establishes institutional mechanisms for co-operation – the High Level Segment[57] and the Prespa Park Management Committee (PPMC). PPMC is a multilateral international legal entity, responsible for ensuring effectiveness in the achievement of the objectives and commitments set out by the Agreement. It is, *inter alia*, responsible also for keeping direct relations with other international players in the field of environmental protection.[58] The Parties accepted to finance implementation of PPMC work plan through regular annual contributions, which also could be financed from other sources.[59]

---

[51] In Pyli, Greece, on 2 February 2010 (OJ of EUL 258/2, 4.10.2011).
[52] Albania has got the candidate status in 2014. Web (24.12.2016): https://ec.europa.eu/neighbourhood-enlargement/countries/detailed-country-information/albania_en.
[53] PRESPA PARK DECLARATION (2000). The Fifth paragraph of the Declaration contains a commitment that the Prespa Park, as the first transboundary area in SEE, shall consist of respective areas around the Prespa Lakes in the three countries, declared as a Ramsar Protected Site.
[54] PRESPA AGREEMENT (2010), Article 1.
[55] Ibid., Article 2.
[56] Which comprise *inter alia* designing and implementing of coherent strategies and plans and programmes, environmental standards and criteria, co-operation in management of the waters of Prespa Park area, exchange of information – in the frameworks established by applicable international law and EU legislation and policies.
[57] That consists of the ministers responsible for environment of the Parties and one representative of EU.
[58] This responsibility comprises establishing connection with the Ohrid Lake Management Committee, as well as contributing to the process of sustainable management of the Extended Drin River basin, contributing to mobilization of the resources of the Parties to the Prespa agreement and the international community (including direct approach to donors), etc.
[59] Article 15.1.

The composition of PPMC is a *sui generis* example, whose features are of specific interest for this chapter. Namely, abandoning all known regional historic and contemporary practices, Parties to this Agreement decided to design a joint body composed not only of the state administration's representatives but also of the representatives of the responsible ministries for environment and a representative of EU, representatives of local communities (one from each municipality in three countries), representatives of significant local NGOs (one from each Party), representatives of institutions managing protected areas (one from each country) and permanent observers of MedWet Initiative (under the Ramsar Convention) and Ohrid Lake Management Committee.[60] Decisions of the Committee shall be taken by consensus, in absense of which the case shall be submitted before the High Level Segment.[61] The Parties accepted to implement and enforce the recommendations of PPMC, and in case of inability, to inform back to the PPMC, explain the reasons for inability for implementation and propose modalities and time frame for implementation.[62]

The Prespa Park Agreement was ratified by Albania, Macedonia and the EU. Ratification of Greece is still pending (slowly entering the eighth year), while local stakeholders and the international community, including MedWet, are urging the Greece Government to ratify the Agreement, and enable full scale co-operation, which is actually limited on the activity at the level of (still) interim (unofficial) PPMC, supported by international funding.[63] In the meantime, Albania and Macedonia, made a new move (identified as a goal to be achieved years ago in Ohrid Agreement 2004), inviting Greece to join them, and establishing a UNESCO transboundary biosphere reserve now in place.[64]

3. Skadar Lake

The Skadar Lake is the largest transboundary water body in South Eastern Europe is of tectonic-carstic origin and originated during the times of the Tertiary or Quaternary age. Its surface area varies between 353 km² (at minimum water level) and 500 km² (at maximum water level), with 335 km² in Montenegro and 165 km² in Albania. There are a number of tributaries and small streams that flow into the lake. Inflow also comes from precipitations, springs and groundwater. The lake outflows through

---

[60] Article 10.3.
[61] Article 9.
[62] Article 12.4. For an in-depth review of the international law aspects of these PPMC characteristics, *see* Bogdanovic, S. (2015).
[63] SPP (2012); SPP (2015).
[64] SPP (2015).

the Bojana/Buna River into the Adriatic Sea, downstream of the outlet of the Skadar/Shkodra Lake into the Buna River.[65] The geographical position of the Skadar/Shkodra Lake can be seen below in Map 5.4.

Due to the high diversity habitats and species, the Skadar/Shkodra Lake is a protected area in both Albania and Montenegro. In Montenegro, it has the status of national park (IUCN category II), since 1983, while in Albania it has been declared as the Managed Natural Reserve (IUCN category IV), since 2005.[66] The lake also has the status of Ramsar and Emerald sites. Scientific efforts are ongoing, under the aegis of IUCN in cooperation with local partners, to develop sub-zones in the protected areas, the objective which would be of ensuring sustainable development in sub/zones.[67]

According to available sources, the Memorandum of Understanding for Co-operation in the Field of Environmental Protection and Sustainable Development Principle Implementation between the Ministry of Environment of the Republic Albania and the Ministry of Environment and Physical Planning of the Republic Montenegro (MoU)[68] has been serving as the legal ground for an active dialogue between the Albanian and Montenegrin governments. By signing this MoU, the two littoral countries committed themselves to conserve the lake's natural resources in a co-ordinated and integrated manner and to improve "the relevant level regulatory and institutional capacities".[69] Signed in 2003, it led to the signing of the Protocol on Co-operation between the University of Shkodra and the University of Montenegro on scientific research of the Skadar/Shkodra Lake.[70] According to the Lake Shkoder Transboundary Diagnostic Analysis,[71] the MoU contains "the basis for co-operation and establishment of a body for water management. This MoU stipulates that working groups will be created and an action plan prepared for its implementation. Although signed in May 2003, no working group has been created yet, let alone an action plan prepared".[72]

It is reasonable to suppose that the MoU (this author was not so lucky to obtain the text of it) also has been serving for other intergovernmental activities regarding construction of the Hydropower Plant "Ashta" (on

---

[65] GEF-WB (2007).
[66] GEF-WB (2007).
[67] IUCN (2016., 05.12.2016.
[68] Signed in Podgorica, Montenegro on 9 May 2003, in the framework of the Stability Pact for SEE project supported by Swiss Development Co-operation (SDC) funds and implemented by Regional Environmental Center (REC) – REC (2019).
[69] Id.
[70] Kalezic, L. (2011). Environmental Impact Assessment in a Transboundary Context in Montep. 226.
[71] ROYAL HASKONING (2006).
[72] Id.

100 *Lake Governance*

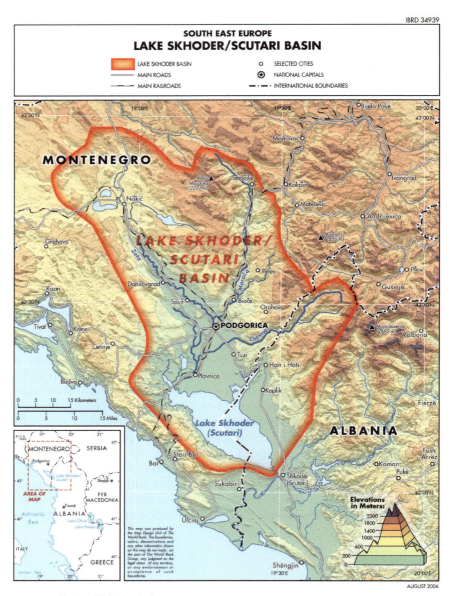

Map 5.4. Skadar/Shkodra Lake.

the Drim River in Albania) and the process of designing and adoption of the Detailed Spatial Plan of Multi-Purpose Accumulation on the Moracha River (in Montenegro).[73] That MoU supposedly makes the same legislative framework for later joint efforts, which led to signing two declaratory (not legally binding; commitment) instruments, the brief review of which follows.

Declaration of the Skadar/Shkodra Lake[74] was signed by the representatives of the bilateral Skadar/Shkodra Lake Forum, composed of a number of stakeholding institutions from both Albanian and Montengrin parts of the Skadar/Shkodra lake catchment area.[75] The Forum built on this Declaration on the joint vision of the Skadar/Shkodra Lake with the Bojana-Buna River as a cross-border protected area[76] and recalls on the Memorandum of Understanding signed by two ministries responsible for environmental issues. By this Declaration, its signatories, *inter alia,* expressed their satisfaction with the work of responsible national institutions on cross-border co-operation aimed at ensuring better protection and at support to achieving sustainable development for the benefit of all members of local communities living in the Skadar/Shkodra Lake region. At the same time, they expressed their commitment to continue cross-border promotion of sustainable development, for the benefit of the local people in the region.

Recognizing the importance of the Shkodra Lake ecosystem (site) and its sustainable management as a contribution towards the halt of the biodiversity loss, the Ministries of Albania and Montenegro responsible for environmental issues[77] signed the Declaration[78] setting the third Saturday of June as the Shkodra Lake Day. Additionally to the commitment aimed at co-operation and taking appropriate measures for celebration of the Lake Day every year, the signatory parties also committed themselves to contribute to the designation of the transboundary Ramsar site for the

---

[73] *Op. cit.*
[74] Shkodra, Albania, 1 February 2006.
[75] Signatories of the Declarations are: University «Luigj Gjurancy», Shkodra, Municipality of Shkodra, Fishery Inspectorate, Project Implementation Unit for GEF/WB project for Skadar/Shkodra Lake, Regional Environmental Agency, Regional Education Directorate, Forestration Directorate (Albania), National Park Skadar Lake, Municipality Podgorica, Institute for the Protection of Nature, Primary School «Savo Pejanović», Ministry of Environment and Physical Planning of the Republic of Montenegro, NGO «Centre for Bird Protection and Research» (Montenegro).
[76] Formulated and adopted by relevant stakeholders from both countries in Bar, Montenegro, in June 2001.
[77] The Ministry of Environment, Forest and Water Administration of the Republic of Albania and the Ministry of Environmental Protection and Physical Planning of the Republic of Montenegro.
[78] Declaration on the Occasion of the Designation of "Lake Day" for Shkodra Lake, Shkodra, Albania, 18 June 2006.

Lake, to the transboundary project implementation with joint structure and to co-operation on establishing of the monitoring system in the Shkodra Lake catchment area in a transboundary pollution context.

4. River Drin Basin: A Strategic Shared Vision

Discussion about a concept for enhanced co-operation among the riparian parties of the Drim River basin ("extended transboundary Drin basin") had began in 2006 between the representatives of the ministries competent for water resources of Albania, Macedonia and Montengro and other stakeholders, in the framework of the Petersberg Phase II/Athens Declaration Process and GEF IW: LEARN Programme in Ohrid, Macedonia. Two years later, based on conclusions from that meeting, a consultative meeting on integrated management of the extended Drin River basin was organized,[79] the outcomes of which was significant in terms of providing the mandate to the Petersberg Phase II/Athens Declaration Process and UNECE to facilitate the establishment of a Shared Vision for coordinated management of the Drin River basin. A year thereafter, the Drin Dialogue, a co-ordinated and structured consultation process, was launched in Podgorica, Montenegro,[80] with the UNECE and GWP-MED empowered to facilitate the Dialogue. The Drin Core Group (DCG), an informal body, was formed at the same event,[81] to serve as a forum for co-ordination among states involved and enabling communication and co-operation between them and the key stakeholders, and as a co-ordinator and facilitator of the Drin Dialogue implementation.[82] In 2010–2011 a project, funded by the Swedish Environmental Agency, supported the Drin Dialogue. All previous activities and the outputs of that project[83] led directly towards signing the Memorandum of Understanding (MoU) for

---

[79] Tirana, Albania, 24.11.2008.
[80] The meeting held on 1.12.2009 was attended by the representatives of the ministries competent for water resources in Albania, Macedonia and Montenegro, as well as by the representatives of the existing lake basin managements committees/commissions (formal or informal).
[81] For more details see: http://www.twrm-med.net/southeastern-europe/supported-processes-and-projects/drin-river-basin/establishing-cooperation-among-the-drin-riparians.
[82] DCG has been composed of representatives of the riparian states, The Prespa Park Management Committee, The Ohrid Lake Watershed Committee, The Skadar/Shkoder Commission, UNECE, GWP-MED and Mediterranean Office for Environment Culture and Sustainable Development (MIO-ECSDE).
[83] Which comprised: identification and analysis of key issues regarding water resources management in the river basin and key national and transboundary stakeholders; functioning of DCG; implementation of the Drin Dialogue through the national and regional consultative meetings; and elaboration and reaching an agreement on the long-term Strategic Shared Vision for the Management of the Drin River Basin.

the management of the Extended Transboundary Drin Basin THE DRIN: A STRATEGIC SHARED VISION.[84]

The objective of the MoU has been formulated as the commitment of the Parties to promote joint action for the co-ordinated integrated management of the shared water resources in the Drin basin, as a means to safeguard and restore ecosystems and the services they provide and to promote sustainable development across the Drin basin.[85]

The Parties of MoU express their commitments to undertake concrete actions to address issues identified as affecting sustainable development in the Drin River basin or in one or more of its sub-basins, in terms of improving access to relevant data and information, providing conditions for sustainable use of water and other natural resources, developing co-operation and measures to minimize flooding, improving management of solid waste, decreasing nutrient pollution and pollution from hazardous substances, etc.[86] Priority actions envisaged to be undertaken at national, bilateral and/or multilateral levels are grouped as short-term (until 2013), medium-term (2015) and long-term (after 2016). By the MoU, the mandate of DCG was expanded to facilitate communication and co-operation among Parties for implementation of MoU. The functions and responsibilities of DCG are elaborated in an Annex to the MoU, while GWP-MED has been empowered to serve as the DCG Secretariat. DCG shall take its decisions by consensus of the representatives of the Parties, which indicates that other members of the DCG (representatives of the lakes communities, commission and international organizations) do not have the right to vote on decisions. DCG meetings should be scheduled at least once a year. The Meeting of the Parties is an institutional arrangement envisaged to meet annually and review the progress in implementation of MoU (Article 6), with power to change set-up, functions and responsibilities of DCG (Annex, ix).

Understanding the need for including the stakeholders views for implementation of the Strategic Shared Vision, the Parties expressed their appreciation and acceptance of the UNECE and GWP-MED to facilitate and organize an annual meeting of the stakeholders in the Drim River basin and encourage the establishment of the "Drin Water Partnership", a mechanism that would engage stakeholders in implementation of the MoU, through facilitating awareness rising, information exchange,

---

[84] MoU was signed by the ministers competent for water management issues and environmental issues of Albania, Macedonia and Montenegro, in Ohrid, Macedonia, on 25.11.2011. For more details see web (16.12.2016): www.twrm-med.net/southeastern-europe/supported-process-and-projects/drin-river-basin/the-memorandum-of-understanding-for-the-management-of-thedrin.
[85] Article 2.
[86] Article 3.

communication, capacity building, consultation and active participation. GWP-MED with MIO-ECSDE are requested to elaborate on such a scheme and establish such a partnership under the auspices of GWP-MED. With regards to funding, the Parties expressed their willingness to ensure participation of their Governments in providing resources for implementation of the MoU, within their abilities, and call upon and invite EU, GEF and other donors to join and provide support in this case. Finally, the Parties requested continuation of the assistance provided under the Petersberg Phase II/Athens Declaration Process, coordinated by the German and Greek Governments and the World Bank, as well as under UNECE Water Convention, and urged UNECE and GWP-MED to continue providing their technical support and facilitation of the process.[87]

The legal effect of this MoU is such that it shall not affect the status of other bilateral relations and rights and obligations of the Parties under earlier MoUs and agreements concluded among them.[88] The MoU became valid from the date of signing.[89] The Parties are allowed to withdraw from the MoU by giving a notice to other Parties, which shall become effective with respect to such a Party 30 days after receipt of such notice by all Parties.[90] The MoU can be terminated by a decision taken by the majority of the Parties.[91]

5. Dojran Lake

Science considers the Dojran Lake, formed in the Neogen-Quaternary,[92] to be a relict of the Plio-Pleistocene Peonic Lake, originated due to the volcanic and tectonic activities.[93] In the past, the lake water area was 42.5 km^2 (27.1 km^2 belonging to Macedonia and 15.8 km^2 to Greece)[94]. It is the smallest of the natural lakes in Macedonia[95] and the warmest lake in Macedonia,[96] due to the proximity of the Aegean Sea. The Dojran Lake was described as one of the European most productive natural lakes, with the annual fish production of 800 tons (1950) to 548 tons (1981).[97] However, in the period 1988–2002, the water level decreased at the historically lowest (recorded)

---

[87] Article 5.
[88] Article 7.
[89] Article 8.
[90] Article 9.
[91] Article 10.
[92] Stojanov, R. et al. (2003).
[93] Francke, A. et al. (2013).
[94] Visual imagery of the Dojran Lake can be found at http://www.paintmaps.com/find_location.php.- Web19.01.2017.
[95] Kostov, V. and Knaap, Martin van der (s.a.).
[96] Web (19.01.2017): http://www.worldlakes.org/lakedetails.asp?lakeid=10447.
[97] Kostov, V. and Knaap, Martin van der (s.a.).

water level of 29.5 km²,[98] due to (uncontrolled) overexploitation of water resources (for irrigation). Decreasing the water level, accompanied with water quality deterioration, resulted in diminishing of biodiversity and plankton reduction. The deterioration of ecosystem harmfully affected the economy in the region[99] (not only fishery, which recorded production of only 25 tons, in 2002, but also tourism and agriculture). An expensive project of the Government of Macedonia has been implemented from 2002, with aim of restoration of the Dojran Lake, by transferring groundwater from the wells in Gjavato Pole through a 20 km long pipeline into the Dojran Lake. Until 2014, an increasing water trend was recorded.[100] However, the water transferred from underground wells was reported as not contributing to the lake's water quality, due to the content of certain pollutants.[101]

For this chapter, there are several specific details, observed and pointed out by researches, indicating the transboundary features of the on-going management/governance of the Dojran Lake. As the main obstacle for Dojran Lake sustainable water resources management, scientists identified the lack of exchange of data and information between Greece and Macedonia, noting that hydrological analysis was only based on the data collected in the period 1951–2010, at the meteorological and hydrological gauging stations located on the Macedonian lake side.[102] Added to this should be a call for co-operation between Greek and Macedonian experts, and specific emphasise of the fact that signing a binding bilateral treaty on the protection of the Dojran Lake between the littoral states is still looked for in future.[103]

Concluding Comments

Co-operation among littoral states (Albania, Greece and SFR Yugoslavia), in earlier times, before the fall of the Berlin Wall, on the Ohrid, Prespa, Skadar and Dojran lakes, based on bilateral agreements aimed at enabling responsible authorities/governments to resolve certain development issues connected to transboundary aspects of shared waters, was

---

[98] Bonacci, O. et al. (2014).
[99] Bonacci, O. et al. (2014).
[100] Id.
[101] Kostov, V. & Knaap, Martin van der (s.a.).
[102] Bonacci, O. et al. (2014). The authors aimed their paper at "stimulation of a close international interdisciplinary monitoring and co-operation", pointing out that "of crucial importance is to facilitate the free and unrestricted exchange of data and information, products and services in real and non-real time on matters relating to the safety and security of society, economic welfare and the protection of Dojran Lake environment".
[103] Stojanov, R. et al. (2003).

weak and random. After the dissolution of SFR Yugoslavia, a necessity evolved for new relations to be established among them. A completely new (environmental/water) management paradigm (new in relation to known "water economy" practice in the region), based and developed in the broader (global, UNECE, EU) legal frameworks established by multilateral treaties applicable in the western Balkans, appears as a powerful driver of change of respective policy, law and practice in the region. Development and implementation of new, "shifted paradigm" which can be denominated as good water/environmental governance appears as a concept strongly supported in a long period of time by the most important international players in the region (UNECE, UNWC, EU MedWet, German, Swiss, Greek and other governments, etc.). The reason for such lasting interest can be seen in the exceptional natural (landscape, biodiversity, ecosystem) and cultural values present in the lakes catchment areas and Extended Drin River basin, which are not only of the interest of littoral/riparian countries, but also are global, European and EU spots worth being preserved, improved and protected.

Continued support of the international community has provided a persuasive and inspiring ambient for establishing relations between the new, post-Yugoslavia, countries of Macedonia and Montenegro and Albania and Greece, and development of transboundary co-operation not only at the level of national authorities, but also between scientific/ research institutions, municipal authorities, NGOs and other stakeholders. A number of important policy (joint commitment-expressing) instruments and necessary strategic planning instruments were developed. Those efforts led to two cases, in signing the binding international legal instruments – the Ohrid Agreement and the Prespa Park Agreement. Noted activities are on-going, and it seems that a clear prospect is ahead, particularly having in view open EU and NATO integrative processes.

There are visible issues in the littoral countries, spanning from weak institutional set-ups, low level of law enforcement, continuous pollution trends, etc. There are also specific issues in bilateral relations between the countries, influencing smooth and rapid achievements of the objectives, which the actors on the western Balkans environmental scene have agreed about. The Dojran Lake, for example, still awaits a strong concerted action of both littoral countries, Greece and Macedonia, for its protection. It seems that the issue of effectiveness of efforts is the one requiring more attention. More attention in view of choosing stronger international binding instruments – political will, i.e., commitments (on objectives, co-operation and means to be applied) expressed in numerous declarations of the authorities are clear, and need now to be translated into practical actions based on binding international treaties, stronger than MoUs are, or declarative, gloomy provisions of agreements. Implementation of all international law regimes applied in the protected areas in all four countries also require more attention. Specific focus on national capabilities for strengthening of enforcement of national environmental/water law is also at stake, etc.

The transboundary co-operation regarding the western Balkans lakes, which enjoys sustained, prospect-opening international support, can and should be seen as a process of change, an integral part of broader changes of littoral countries and their societies, and their relations with neighbours, in a world changing substantially after the fall of Berlin Wall, dissolution of Yugoslavia and wars that came thereafter, and also in the changing Nature (caused by climate change and rising pressure of human activities). The huge continuous specific working ambient in this (sub)region, aimed at coping with changing ambient and streaming towards sustainable solutions, has led a clear awareness (on all layers – scientific, administrative, municipal, NGO, etc.) of the necessity of practical implementation of a new multilayer governance, which could bring all stakeholders in the lakes basins (meaning of course, in the Extended Drin River basin), for transboundary negotiating and decision-making round table, ensuring them all an equal participating voice in the decision-making process on the environment they live in. In that sense, the institutional set-up introduced by Prespa Park Agreement is a *sui generis* example of good governance practice, having a serious potential, once when it is implemented, to become a paradigmatic model for successful peaceful co-operation. Hopefully, the Greek Government shall find the way to ratify it in spite of the lasting pitfalls two countries found themselves in regarding the official name of Macedonia. It seems unavoidable, sooner or later. As soon as better for all.

Acknowledgements

The author wishes to express his gratitude to Dr. Lillian Castillo de Laborde, Professor of International Law at the Buenos Aires University, for her valuable comments, which improved this paper.

Credit goes to the Ministry for Sustainable Development and Tourism of Montenegro for providing several important documents.

Credit also go to the World Bank Group for providing and permission for reproduction in this book of the maps of the Extended Drin River Basin, the Ohrid Lake basin, the Prespa Lakes Basin and the Skadar/Shkodra Lake Basin.

References

Athens Declaration. (2003). TWRM in SEE: Athens Declaration (2003); Web (25.12.20126): http://www.twrm-med.net/southeastern-europe/regional-dialogue/framework/petersberg-phase-ii-athens-declaration-process/athens-declaration/English.pdf.

Avramoski, O., Kycyku, S., Naumoski, T., Panovski, D., Puka, V. and Selfo, L. (2014). Lake Ohrid Experience and Lessons Learned Brief; http://projects.inweh.unu.edu/inweh/display.php?ID=4228 ; Web, 17.10.2014.

Birnie, P. and Boyle, A. (2002). International Law & the Environment; Oxford University Press, New York, 2002.

Bogdanovic, S. (1993). Legal aspects of danube waters protection. *Acta Juridica Hungarica*, Budapest, 35: (3-4): 321–332.

Bogdanovic, Slavko. (2005). Legal aspects of transboundary water management in the danube basin. Large Rivers. *Arch. Hydrobiol. Suppl.*, 158/1-2, 16(1-2): 81–82.

Bogdanovic, Slavko. (2008). Prespa Park Co-Ordination Committee in Transboundary Ecosystem Management; Final Technical Assessment Report, prepared in the framework of the UNDP project Integrated Ecosystem Management in the Prespa Lakes Basin of Albania, Fyr Macedonia and Greece, Novi Sad, Web (24.01.2017): http://iwlearn.net/iw-projects/1537/reports/integrated-ecosystem-management-in-the-prespa-lakes-basin.

Bogdanovic, Slavko. (2015). The concept of state sovereignty and river/lake basin law in South Eastern Europe. pp. 537–558. *In*: Tvedt, T., McIntire, O. and Woldesadek, T.K. (eds.). A History of Water, Series III, Volume 2: Sovereignty and Water Law; I.B. Tauris, London-New York.

Bonacci, Ognjen, Popovska, Cvetanka and Geshovska, Violeta. (2014). Analysis of transboundary Dojran Lake mean annual water level changes; DOI 10.1007/s12665-014-3618-6.

CSCE/OSCE (1975). CSCE (later OSCE): Conference on Security and Co-Operation in Europe Final Act (Helsinki 1975).

EU (2016); European Commission: European Neghbouring Policy and Enlargement; Web (26.01.2017): https://ec.europa.eu/neighbourhood-enlargement/countries/detailed-country-information/montenegro_en.

Fernandez-Jáuregui, Carlos and Crespo Millet, Alberto. (2008). Water a unique resource. *In*: Mancisidor, M. and Uribe, N. (eds.). The Human Right to Water – Curent Situation and Future Challenges. UNESCO Etchea—Centro UNESCO del Pais Vasco; Barcelona.

Francke, A., Wagner, B., Leng, M.J. and Rethemeyer, J. (2013). A Late Glacial to Holocene Record of Environmental Change from Lake Dojran (Macedonia, Greece); doi: 10.5194/cp-9481-2013.

Gardin, Jean. (2007). The Tri-National Prespa Park in Albania, Greece and Macedonia (FYROM): Using Environment to Define the New Boundaries of the European Union. Borders of the European Union: Strategies of Crossing and Resistance, 2007. HAL ID: halshs-01170554; Web (24.12.2016): https://halshs.archives-ouvertes.fr/halshs-01170554.

GEF-WB. (2007). Global Environmental Facility (GEF) & World Bank (WB): The Strategic Action Plan (SAP) for Skadar/Shkodra Lake Albania & Montenegro; Developed in the framework of the Lake Skadar/Shkodra Integrated Ecosystem Management Project.

Government of Montenegro. (2016). National Strategy for Transposition, Implementation and Enforcement of EU acquis on Environmental and Climate Change for the period 2016–2020; Podgorica, July 2016. Web (19.01.2017): http://www.mrt.gov.me/en/library/strategije.

IUCN. (2014). IUCN: Supportive Effective Management of Lake Ohrid. Web (22.01.2017): https://www.iucn.org/content/supporting.effective-management-lake-ohrid.

IUCN. (2016). IUCN: A boost to cross-border conservation between Albania and Montenegro; Web: (22.01.2017): https://www.iucn.org/news/new-zoning-plan-lake-skadar.

Kalezic, Lazarela. (2011). Environmental impact assessment in a transboundary context in Montenegro. pp. 213–227. *In*: NATO.

Kostov, Vasil and Knaap, Martin van der. (s.a.). The collapse of fisheries of Lake Dojran// reasons, actual situation and perspectives; conference proceedings—IV International Conference FISHERY, *s.l., s. a.*

Mici, Ardiana. (2016). Governance of Natural Resources in Transboundary Freshwater Settings – The Ohrid-Prespa Transboundary Biosphere Reserve (Power Point presentation at Act4Drin Spring School 07–15 May, 2016). Web (22.12.2016): http://mio-ecsde.org/wp-content/uploads/2016/05/Ardiana-Mici_Governance-of-natural-resources.pdf.

NATO. (2011). Montini, M. and Bogdanovic, S. (eds.). Environmental Security in South-Eastern Europe—International Agreements and their Implementation; Springer, NATO Science for Peace and Security Series C: Environmental Security; Dordrecht.

Ohrid Agreement. (2004). Agreement between the Council of Ministers of the Republic of Albania and the Government of the Republic of Macedonia for the Protection and Sustainable Development of Lake Ohrid and its Watershed; Web (17.12.2016): http://faolex.fao.org/docs/pdf/bi-69075E.pdf.

Petersberg Declaration. (1998). Petersberg Declaration; International Dialogue Forum (1998); Web (25.12.2016): http://www.twrm-med.net/southeastern-europe/regional-dialogue/framework/petersberg-phase-ii-athens-declaration-process/petersberg-process/Petersberg%20declaration.pdf.

PMs Joint Statemet. (2009). Joint Statement; Pyli Prespa (27 November 2009).

Prespa Agreement. (2010). Agreement on the Protection and Sustainable Development of the Prespa Park Area; Web: 23.12.2016: http://ec.europa.eu/world/agreements/prepareCreateTreatiesWorkspace/treatiesGeneralData.do?step=0&redirect=true&treatyId=9102.

Prespa Park Declaration. (2000). Declaration on the creation of the prespa park and the environmental protection and sustainable development of the prespa lakes and surroundings. *Agios Germanos*.

REC. (2010). Regional Environmental Center Albania: Assessment on Current Situation of Shkodra/Skadar Lake Ramsar Site, 2010.

Royal Haskoning. (2006). Royal Haskoning: Lake Shkoder Transboundary Diagnostic Analysis Albania & Montenegro; The Final Report at the same title project funded by World Bank (IBRD).

Skarbøvik, E., Nagothu, S.U., Mukaetov, D., Perovic, A., Shumka, S. and Borgvang, S.A. (2008). Transboundary Lakes in Balkan Area, Monitoring and Management in Accordance with the EC Water Framework Directive; BALWOIS 2008 Conference on Water Observation and Information System for Decision Support, 27/31 May 2008, Ohrid; Web (26.01.2017): http://balwois.com/?s=2008+monitoring&post_type=k9proceeding.

SPP. (2012). Society for the Protection of Prespa: Press Release, 09.08.2012; Web (13.09.2016): http://www.spp.gr/index.php?option=com_content&view=article&catid=6%3A2010-03-04-13-52-03&id=113%3A2012-08-08-06-32-42&Itemid=5&lang=en.

SSP. (2015). Society for the Protection of Prespa: Press Release, 02.02.2015; Web (13.09.2016): http://www.spp.gr/index.php?option=com_content&view=article&catid=6%3A2010-03-04-13-52-03&id=172%3A2015-02-02-09-29-18&Itemid=5&lang=en.

SSP. (2017). Society for the Protection of Prespa: Prespa Area; Web (24.01.2017): http://www.spp.gr/index.php?option=com_content&view=article&id=4&Itemid=4&lang=en#5.

Stojanov, Risto, Trcek, Branka, Dolenc, Tadej, Dimkovski, Trajan and Pirc, Simon. (2003). Environmental protection of the Dojran lake catchment area; *RMZ—Materials and Geoenvironment*, 50(1): 369–372.

Tasevska, Orhideja, Kostoski, Goce, Guseska, Dafina, Patcheva, Suzana and Trajanovski, Saso. (2016). PA: Ohrid and Prespa Lakes; ECO Potential General Meeting, 27–30 April, 2016, Texel., The Netherlands; Web (22.01.2017): http://ecopotential-project.eu/images/ecopotential/protected-areas/Ohrid.pdf.

TWRM. (2012). A Regional Dialogue for TWRM in See: Establishing Cooperation Among the Drin Riparians; Web (22.01.2017): http://www.twrm-med.net/southeastern-europe; and http://www.twrm-med.net/southeastern-europe/supported-processes-and-projects/drin-river-basin/establishing-cooperation-among-the-drin-riparians.

UNESCO WHC. (2016). UNESCO World Heritage Centre (UNECSO WHC): 4th Transboundary Platform meeting for the safeguarding of the Lake Ohrid Region; Web (22.12.2016): http://whc.unesco.org/pg_friendly_print.cfm?cid=83&id=1339&.

CHAPTER 6

Governance of Transboundary Water Commissions

Comparison of Operationalizing the Ecosystem Approach in the North American Great Lakes and the Baltic Sea

Savitri Jetoo-Åbo * and *Marko Joas-Åbo*

Introduction

The Laurentian Great Lakes and the Baltic Sea are two large transboundary water systems in North America and Europe respectively. Despite the geopolitical, geographical and ecological differences, these water bodies share a similar history, with each signing transboundary governance agreement in the early 1970s due to concerns of water pollution. In 1972, the Great Lakes Water Quality Agreement was signed between Canada and the United States of America amidst concerns about the 'dying' Lake Erie. This led to the new role under the Agreement for the International Joint Commission (IJC), the transboundary institution tasked with oversight of this agreement. At the same time in Europe, during a period of détente in the Cold War era enhancing a revival of a state centred international organization system (with for example the creation of OSSE), negotiations

---

Åbo Akademi University, Vanrikinkatu 3, Turku, 20500, Finland.
Email: marko.joas@abo.fi
* Corresponding author: savitri.jetoo@abo.fi

were ongoing for a transboundary agreement amidst concerns about increasing pollution to the Baltic Seas. This culminated in a transboundary water agreement in 1974, the Helsinki Convention, which established the Helsinki Commission (HELCOM) as the coordinating body. Whilst these commissions can be seen as successes because they were effective in bringing the key national players to the table, the continued degradation of these transboundary water ecosystems would suggest that they are not successful in applying the ecosystem approach to governance, as called for in both transboundary agreements. The ecosystem approach is a useful approach for transboundary water governance as it looks beyond the political boundaries and more at natural boundaries, offering an integrated approach for the management of land, water and living organisms. This chapter investigates the effectiveness of these transboundary water commissions in operationalizing the ecosystem based approach to governance by assessing their adaptive capacity, the governance capacity for dealing with change. It starts by looking at the evolution of these commissions, it uses a framework for adaptive capacity from the literature and assesses the performance of these transboundary commissions against these principles. It then identifies gaps and makes recommendations that can inform policymakers.

The Ecosystem Approach and Adaptive Capacity

As a normative and scientific concept, the ecosystem approach to management can be traced back to the better understanding for the interplay between nature and human society as well as to the early ecosystem science in the 1960s and 1970s (Szaro et al. 1998). The ecosystem approach was introduced as a concept under the 1978 Great Lakes Water Quality Agreement, the first revision of the 1972 Great Lakes Water Quality Agreement. Whilst the word 'ecosystem' was used 34 times in the 1978 Great Lakes Water Quality Agreement, the ecosystem approach debuted once in Annex 2 under 'Remedial Action Plans and Lakewide Management Plans' (US and Canada 1978). This was the first transboundary agreement to recognize the importance of the ecosystem approach for restoring ecosystem services (beneficial uses). This approach was made more formal and international when it was adopted in the Convention on Biological Diversity (CBD 1995). It was further clarified as the management of land, water, living resources and humans as interconnected systems (COP5 2000), termed as social-ecological systems (Berkes and Folke 1998). The concept has gradually been introduced to a growing number of national, regional and international governing instruments for human – environment interaction. Based on the use of the concept as well as academic literature, five core principles can be seen to characterize the ecosystem approach for management: (a) an inclusive holistic approach, (b) scale dependent management and integration, (c) sound knowledge based science, (d) broad participation, and finally (e) adaptive management including ecosystem services (Söderström and Kern 2017).

The Helsinki Commission (HELCOM 2017) has adopted the ecosystem approach in the Baltic Sea Action Plan in 2007. It defines that "the ecosystem approach is based on an integrated management of all human activities impacting on the marine environment and, based on best available scientific knowledge about the ecosystem and its dynamics, identifies and leads to actions improving the health of the marine ecosystem thus supporting sustainable use of ecosystem goods and services". Since the ecosystem is at the centre of the ecosystem approach, and is highly dynamic and complex with incomplete understanding of their functioning, management needs to be highly adaptive, containing feedback elements (COP5 2000).

Due to the non-linear, uncertain nature of ecosystems, traditional command and control systems need to be replaced by more inclusive governance that can lead to legitimate outcomes (Aswani et al. 2012). Such adaptive governance has been described as environmental governance regimes that can deal with adaptability and change (Olsson et al. 2006). Adaptive governance systems facilitate adaptive capacity, which can be defined as the "ability or potential of a system to respond successfully to change" (IPCC 2007). According to Pahl-Wostl (2009), adaptive capacity can be defined as "the ability of a resource governance process to first alter processes and if required convert structured elements as a response to experienced or expected changes in the societal or natural environment". As such, adaptive capacity is necessary for transboundary organizations such as the IJC and HELCOM to effectively respond to environmental stressors such as eutrophication, the biggest threat in both areas after climate change.

As such, this study aims to assess the adaptive capacity of both transboundary commissions by using the framework developed by Jetoo and Krantzberg (2016) to determine the presence of adaptive capacity determinants in the functioning of both commissions. The study asks the question, how well are these determinants of adaptive capacity represented in the work of the International Joint Commission and the Helsinki Commission? A second question follows from this, what can each commission learn from the other to increase their adaptive capacity to change?

Formation of the Commissions

The International Joint Commission

The International Joint Commission was founded by the Boundary Waters Treaty of 1909 (the Treaty). While the main driver for this treaty was the concern over apportionment of water for producing hydropower near the end of the 19th century (Dreisziger 1981), the Treaty has a broader scope that includes the prevention of disputes regarding the use of all shared waters, the resolving of differences of "rights, obligations or interests of either in relation to the other or to the inhabitants of the other, along their common frontier" and to settling of any such differences that may arise at a later date (Boundary Waters Treaty Act 1909). It was article III of this treaty that introduced the International Joint Commission as follows:

> "It is agreed that, in addition to the uses, obstructions, and diversions heretofore permitted or hereafter provided for by special agreement between the Parties hereto, no further or other uses or obstructions or diversions, whether temporary or permanent, of boundary waters on either side of the line, affecting the natural level or flow of boundary waters on the other side of the line shall be made except by authority of the United States or the Dominion of Canada within their respective jurisdictions and with the approval, as hereinafter provided, of a joint commission, to be known as the International Joint Commission."

Further, Article X of the Treaty goes on to specify that the IJC can investigate a specific transboundary issue, under a formal request by both governments (worked out bilaterally), termed a 'reference'. Using this provision, the US and Canada issued a joint reference in 1964, to the International Joint Commission (IJC) to investigate pollution of Lake Erie and elsewhere on the lower lakes, perhaps as a result of the growing public and scientific concern about water pollution in North America after World War II (Botts and Muldoon 2005). This active public demanded action, which the IJC countered with public hearings on a summary report of the 1964 reference, followed by a final report and finally, after six years of study and intense negotiations, The Great Lakes Water Quality Agreement was signed by Prime Minister Trudeau and President Nixon on April 15, 1972 (Botts and Muldoon 2005). The initial scope was targeted to limiting phosphorus inputs in order to control algae growth, which gave one of the Great Lakes, Lake Erie, the status of 'dying'. This agreement was later renewed in 1978, 1987 and in 2012 (as the Great Lakes Water Quality Protocol 2012) with the purpose "to restore and maintain the chemical, physical and biological integrity of the Waters of the Great Lakes" (US and Canada 2012).

The 1972 Great Lakes Water Quality Agreement (the Agreement) gave the IJC new responsibilities on water quality; to obtain and interpret water quality data and provide advice and recommendations to the US and Canada (the Parties) on measures to attain water quality objectives. The structure established by the Agreement (Fig. 6.1) included a Great Lakes Water Quality Board (membership by federal, state and provincial governmental agencies) to be the principal policy advisor to the IJC and the Research Advisory Board (membership by research managers) to advise on scientific matters. There have been changes to the organizational structure of the IJC as the Agreement was updated, but the IJC has retained its core function to evaluate the progress of the Parties and make recommendations to further the purpose of the agreement. The Great Lakes Water Quality Agreement Protocol 2012 (the Protocol) maintains the Great Lakes Water Quality Board (WQB) as the principal policy advisor to the IJC, helping with the review and assessment of progress of implementation of the Protocol, with the identification of challenges and emerging issues and also helping with strategic governance advice (such as the role of jurisdictions in implementing the strategies). In the Protocol, the Great Lakes Science Advisory Board (formerly the Research Advisory Board) is given the goal of advising the IJC and the WQB on research and scientific matters. The organizational structure also includes the Great Lakes Regional Office, with the mandate of providing administrative support to the IJC and its Boards.

Figure 6.1. Governance structure of the IJC (Jetoo and Krantzberg 2015) and HELCOM (2017).

The Helsinki Commission

The Helsinki Commission (HELCOM) was established by a transboundary water agreement, the Convention on the Protection of the Marine Environment of the Baltic Sea Area (also known as the Helsinki Convention) in 1974. Whilst the main driver of the Helsinki convention was also concern for pollution (the Baltic Sea was deemed the most polluted sea in the world), it was drafted in very different circumstances than the Great Lakes Water Quality Agreement, as it was negotiated during the Cold War when the Baltic Sea was still very divided by the Iron Curtain. The window of opportunity for the convention appeared during an era of détente in the relations between East and West in the mid-1970s, with, for example, the initiation of the European security organization OSSE, also in Helsinki in 1975. It must be noted, however, that the initiation process also for Helsinki Convention and further HELCOM was a highly state centred one. The Helsinki Convention was the first multilateral water agreement signed by two mutually opposing military alliances; three of the original signatories were members of the Warsaw Pact (the Soviet Union, Poland and the German Democratic Republic), two were members of the North Atlantic Treaty Organization (NATO) (Germany and Denmark), whilst the remaining two signatories defined themselves as neutral (Finland and Sweden) (Räsänen and Laakkonen 2008). It was most expansive in scope, almost encompassing all known pollutants at the time and covering the sea in its entirety (Ehlers 1994). Pollution was defined in Article 2 as follows (Helsinki Convention):

> ""Pollution" means introduction by man, directly or indirectly, of substances or energy into the marine environment, including estuaries, resulting in such deleterious effects as hazard to human health, harm to living resources and marine life, hindrance to legitimate uses of the sea including fishing, impairment of the quality for use of sea water, and reduction of amenities."

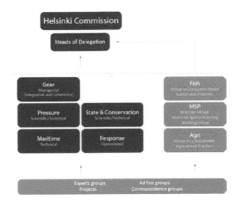

The main obligation of the Contracting parties follows from this in Article 3, either individually or collectively to take all appropriate measures for pollution prevention or abatement and to protect and enhance the marine environment of the Baltic Sea Area.

The Helsinki Commission (HELCOM) was established for the purposes of the Convention in Article 12, with duties delineated in Article 13. Such duties include observing the implementation of the Convention, making recommendations on the measures relating to pollution control including defining pollution control criteria and objectives, reviewing and recommending changes to the Convention, receiving, processing and disseminating scientific information and promoting scientific and technological research. Whilst the IJC is supported by Boards, HELCOM (Fig. 6.1) is given the task under its functions to seek out competent organizations to collaborate on activities that can further the achievement of the purposes of the Convention. The current members of HELCOM, besides the original signatories or their successors (Russia, Estonia, Lithuania and Latvia) also include the European Union (EU), in addition to the nine coastal states. In fact, with the exception of Russia, the Baltic Sea can be considered as an internal EU sea, a fact that has institutional effects for the governance of the Baltic Sea.

The Convention stipulates that the office (Secretariat) be located in Helsinki and that the Commission shall appoint an Executive Secretary, who is to be the chief administrative official of the commission to perform all administrative processes of the commission. Whilst it does not specify which other support personnel are needed, there is a provision for the commission's appointment of other support personnel as necessary. The success of HELCOM now hinges on the effective implementation of the Baltic Sea Action Plan (BSAP), a plan aimed at "restoring the good ecological status of the Baltic Marine Environment by 2021" (HELCOM 2017). This science based plan was adopted by the HELCOM members and EU in 2007 and includes innovative management approaches to aid policy implementation.

Methodology

Adaptive Capacity

This study uses the framework developed by Jetoo and Krantzberg (2016) to assess the adaptive capacity of the transboundary water governance commissions to effectively respond to change. The adaptive capacity framework is comprised of six determinants as follows: **public participation, science, networks, leadership, flexibility and resources (see Table 6.1).**

Table 6.1. Determinants of adaptive capacity (modified from Jetoo and Krantzberg 2016).

Determinant	Description
D1 Public Participation Representation	Adaptive capacity will be built when there is participation of diverse, interested stakeholders to allow access to new modes of knowledge and stakeholder buy in to deal with the highly complex and uncertain environment of eutrophication. One challenge is determining which of the public is included and excluded for efficiency in processes.
D2 Science	Adaptive capacity will be built when sound science is used to guide decision making processes on eutrophication issues across multiple scales.
D3 Networks	Adaptive capacity will be built when different actors operate across multiple scales on the same issues in horizontal governance networks such as epistemic communities, transnational advocacy coalitions and global civil society.
D4 Leadership	Adaptive capacity will be built when there is a new kind of leader (individual, institutional) who can interact with stakeholders and facilitate public learning to overcome uncertainty, distrust and conflict in the highly uncertain environment of eutrophication.
Experience	A leader with more experience would more ably deal with uncertain events in an effective and timely manner.
Decision Making	Complex uncertain problems such as eutrophication require leaders to consider and balance their thinking with others and to engage in new approaches to decision making.
D5 Flexibility	Adaptive capacity will be built when there are mechanisms for information feedback loops that are as a result of monitoring and are used to guide decisions and adjust programs. Adaptive capacity will be built when the legislation and institutions are more flexible.
D6 Resources	Adaptive capacity will be built when there is availability of skilled human resources for functions such as innovation and monitoring and financial resources for implementation of policy measures.

Comparison of Determinants of Adaptive Capacity: IJC and HELCOM

Public Participation

The IJC has had a long history of interacting with the public, with citizens attending as observers to the annual meetings (under the 1972 agreement), to a public information service called for in the 1978 Agreement, to the five non governmental observers participating in the final negotiations of the 1987 Protocol (Botts and Muldoon 2005). This active citizen engagement has continued through associations such as the Alliance for the Great Lakes (https://greatlakes.org) and the Sierra Club. The International

Joint Commission is mandated by the Great Lakes Water Quality Protocol 2012 (the Protocol) (US and Canada 2012) to involve the public, e.g., under Article 7 sub clauses g and h mandates that the IJC consults on a regular basis with the public on issues for ecosystem protection and restoration and engages the public on the value of the Great Lakes. The IJC is also required to provide the public with the Parties (US and Canada) triennial report on Progress and include a summary of the public input in their assessment of progress by the parties. This has led the Parties to the agreement (Canada and the US) to establish the 'Great Lakes Public Forum', a triennial forum open to the public and consisting of presentations on the progress report and input by the public (http://www.participateijc.org/great-lakes-public-forum). These measures allows for the public's assessment of programs and measures implemented under the Protocol (Jetoo and Krantzberg 2014). The IJC commitment to engaging the public can be seen from the organizational structure (under the Protocol), with the representatives of the public having seats on the IJC's Water Quality Board and the Science Advisory Board (Fig. 6.1). It is clear from the IJC's consultant report that there is need for better public participation and collaboration among the public and private sector "to enhance water stewardship by fixing leaking public water infrastructure, supporting innovation, and increasing funding to close the region's water infrastructure deficit, unlock water conservation potential and encourage a water stewardship focus region wide" (Pentland 2015).

In contrast, there is no permanent seat on the organizational structure of HELCOM for members of the public (Fig. 6.1). This can be traced to the Helsinki Convention, as there are stipulations for HELCOM to engage the topic in the list of duties under Article 20 (HELCOM 2017), unlike the case of the IJC. Members of the public can attend HELCOM meetings as observers, but the formal application process limits the full participation of any member of the public as in the case of IJC's public forum. As such, the public have a more limited role in decision making and assessment of progress to measures in implementation of the Helsinki Convention, than the case of the IJC. This can also be seen as a trace from the introduction HELCOM during the Cold War era. The process was fully state centred multilateral process with a limited space for public participation as this was highly limited in general in the original at that time non-democratic member states (the Soviet Union with satellite states). Even today it can be argued that full public engagement would be more challenging for HELCOM with 10 full members, including nine member states (Denmark, Estonia, Finland, Germany, Latvia, Lithuania, Poland, Russia and Sweden) and the European Union, representing millions of persons speaking different national languages, more can be done to give the public a permanent seat at the table. In order to surmount this challenge, HELCOM has organized annual stakeholder meetings for members of the public, where they can comment on HELCOM plans and give input to undertakings such as the Baltic Sea Action Plan. A record of past meetings can be found on the HELCOM website and members of the public can sign up for these meetings via this portal. However, for such a large gathering, it is doubtful that all members of the public would be able to contribute meaningfully. For example, at the seminar HELCOM conducted at the 7th Strategy Forum of the EUSBSR, there were over 100 members of the public present, but the meeting was dominated by a panel discussion and only less than five comments from members of the public (author 1 notes from seminar attendance).

Science

The standard of joint fact finding and research was carried over from the Boundary Waters Treaty into the Protocol, which is standing reference under the treaty and underpins the IJC's functions under the Protocol. This scientific fact finding by the IJC is conducted by the Water Quality Board (WQB) and the Science Advisory Board (SAB), which are groups of experts. Under Article 7 (clause e) of the Protocol (US and Canada 2012), the IJC is tasked with:

> "Assisting in and advising on scientific matters related to the Great Lakes Basin Ecosystem, including: (i) identifying objectives for scientific activities; and (ii) tendering scientific advice and recommendations to the Parties and to State and Provincial Governments, Tribal Governments, First Nations, Métis, Municipal Governments, watershed management agencies, other local public agencies and the Public."

There is a record of achievement by the IJC under this clause, as in the past, such fact finding efforts led to identifying the effects of exposure to contaminants on reproductive consequences for wildlife (Botts and Muldoon 2005). The IJC's commitment to science has enhanced its credibility and has enabled the commission to identify emerging threats, often reported on in special reports such as chemicals of emerging concern. Research conducted by the IJC and its boards helps in identifying problems not previously identified, such as toxic contaminants and can be a basis for global action (as in the case of the Stockholm Convention for Persistent Organic Pollutants). The continued use of science in its decision-making is evident from IJC's meeting minutes, where reports were reviewed on the impact of air pollution and human health outcomes in selected regions in the Great Lakes, on indicators for chemicals of mutual concern and aquatic invasive species (IJC 2016).

Similarly, HELCOM's recommendations are based on sound scientific research and investigations. The preamble to the Baltic Sea Action Plan captures this sentiment, stating:

"HELCOM's monitoring and assessment programme will contribute to an improved scientific understanding of the marine environment that will in turn contribute to the periodic review of the objectives, associated targets and indicators, and will be decisive when determining the need for further management measures" (HELCOM 2017).

This commitment to science is clearly reflected in the BSAP's load reductions requirements for phosphorus and nitrogen which are to be cut by 42 percent and 18 percent respectively by 2021 from the 1997–2003 average loadings and by allocating nutrient input reductions for each member country. These targets were set using the MARE NEST models Swedish Baltic Nest Institute MARE research program system Nest for simulations of annual loadings and connections in the Baltic ecosystem (Wuff et al. 2013). However, there was a strong emphasis on the modelling and input of natural science information, but no emphasis on economic information of member countries. This can retard implementation of the BSAP, as poorer countries would have greater difficulty allocating resources for the load reductions or implementation of the measures may have adverse socio-economic impacts (Haila et al. 2008). However, binding regulations through EU and supportive funding mechanisms do support a common level of intervention.

Networks

The requirement that the IJC provides information to the public and that it utilizes the WQB and the SAB (formerly the research advisory board) helped to create a sense of community with strong networking around Great Lakes issues (Botts and Muldoon 2005). This has led to a very networked multi-level governance structure for the Great Lakes, with members of the federal, state, provincial and municipal governments and non-governmental organizations given seats at the table of these boards. Through their positions on these boards, members can contribute to the IJC as neutral members and influence their organizations when they return from board meetings. The Great Lakes Water Quality Protocol also calls for the governments of Canada and the United States to establish a Great Lakes Executive Committee (GLEC) (https://binational.net/glec-cegl/), comprising of key federal and municipal agencies, to collaborate on implementation of measures in the Protocol. The Great Lakes Executive Committee functions as a coordinating body in the multilevel governance landscape, bringing together federal, provincial, state, municipal, non-governmental and private citizen actors in the Great Lakes governance landscape.

The provisions in the Protocol for public engagement should in theory also lead to strong networking. The Protocol is more expansive as it includes provisions for indigenous representation on boards and in decision making forums. At the IJC's public forum, there was representation by members of the Lake Waterkeepers groups, by Great Lakes Alliance, by epistemic communities and by the general public. However, to prevent the recurrence of a disengaged Great Lakes Community, which occurred in the 1980s and 1990s when the IJC changed the format of the meetings to limit public participation due to pressure from the Parties (Botts and Muldoon 2005), the parties need to remain committed to considering the recommendations provided by the IJC, even if they are unpopular with segments of the electorate such as industry or agricultural interests. Recommendations of 'zero discharge' of toxic contaminants were unpopular with the parties, and chemical industry, who started to attend the IJC meetings (Botts and Muldoon 2005).

The (policy) network structure in the Baltic Sea region exploded after 1991 with fall of the Iron Curtain and the multi-level governance development in the region and in the EU. In executing its mandate under the Helsinki Convention, HELCOM has organized stakeholder meetings for members of the public, including organized interests. This aids in networking among stakeholders including international organizations, subnational actors, epistemic communities, private sector and other non governmental organizations, creating a networked environmental regime for Baltic Sea governance. This diversity of organizations may be linked to HELCOM's record of advocating tougher regional environmental policies, attracting policy advocates who seek to influence the outcomes (VanDeveer 2011). Some of the international organizations that are part of HELCOM's networks as observers include the European Boating Association, European Chlor-Alkali Industry, European Community Shipowners' Association, Fertilizers Europe, Federation of European Aquaculture Producers, Federation of European Private Port Operators and Federation of European Aquaculture Producers (HELCOM 2017c). The extent of HELCOM networking can be seen with observers are far as the Great Lakes, with the Great Lakes Commission officially registered as an observer. Whilst there is clearly a diversity of actors that network with HELCOM, there is no clear indication how their views are taken into consideration in HELCOM's recommendations to its members on Baltic Sea ecosystem restoration. Efficient networking has the potential to lead to more influence on member countries implementation measures, with HELCOM acting as a glue that binds different organizations in the region to work together. More efficient networking would mean more effective communication, leading to flexible adaptive actions in times of change.

Leadership

The operations of the IJC were historically characterized by binationalism (personal interests set aside for common goal), enabling it to function as a politically independent agency with three commissioners from the US and three from Canada who made impartial decisions in the interests of the lakes. This allowed members of the expert advisory boards to operate without home agency constraints, leading to consensus in decision making, as the ecosystem restoration was the common goal. The experience and early leadership of the IJC was enabled by the terms of the agreements that led to a culture of accountability and openness, measures such as reporting on progress (biennial and now triennial reports by the Parties), the provision of information to the public and allowing the public to attend meetings. IJC was seen as an expert leader, as decision making was based on scientific evidence. It was the impartial political leadership of IJC that led to the early success under the agreement, but this was later undermined with political appointments of commissioners to the IJC with the change of administration (US) and government (Canada) (Botts and Muldoon 2005). Political appointments can lead in gaps in membership to the commission and this can lead to a lack of continuity and eroding of leadership by the commission. Changes in member appointments and mandates might change the working climate in the IJC, for example, currently, there are two US commissioners on the IJC website with no explanation on the absence of the third; this has the potential to weaken the leadership provided by the IJC.

HELCOM is clearly at the centre of Baltic Sea governance and as such, an experienced leader in organizing stakeholders and in providing scientific information, as is seen by the many reports published and available on their website. According to HELCOM, conflicts within working groups and member countries can be resolved through mediation by a third party (a contracting party, qualified person or qualified international organization), and if it escalates, it can go to a tribunal or the International Court of Justice. These established conflict resolution procedures helps HELCOM in steering member activities and enhance its role of leader. Since each member state has a vote and decision making is by consensus, HELCOM is institutionally enabled to function as an effective decision maker.

The fact that almost all HELCOM member states are today EU members, and that EU is also an institutional member in HELCOM, provides both possibilities and obstacles for political leadership in the region. EU has a legislative mandate over member states, in strength varying by sector, but finding a common view already on the Union level makes decision making easier within HELCOM. As such, EU has also highlighted the region in the first macro-regional strategy for EU: EU Regional Strategy for the Baltic Sea Region (EUSBSR), adopted in 2009. The downside in this comes with the position of Russia as the only non-EU member in HELCOM. In a politically tense climate, international organizations as HELCOM easily become political weapons over hegemony.

Whilst HELCOM is clearly seen as a scientific leader, there was a call for more transparency and political leadership by HELCOM by members of public during a HELCOM seminar held at the 7th Strategy Forum of the EUSBSR (HELCOM 2017d), with specific mention of lack of clarity of documents on HELCOM website.

Flexibility

Monitoring and feedback of information is required for learning loops necessary to build adaptive capacity for governing in times of change. The research carried out by the IJC and its boards has in the past lent flexibility to the functions of the IJC, as it led to the development and acceptance of new ideas. It was research that led to the pinpointing of phosphorus as the limiting nutrient for eutrophication in Lake Erie in the 1970s (Schindler et al. 1971). The discovery of toxic contaminants led to the articulation of the need for an ecosystem approach in the 1978 agreement (Botts and Muldoon 2005). With its oversight role of government efforts in monitoring and data collection in the Protocol, the IJC is equipped to continue its efforts on policy modifications and innovations recommendations to the Parties. There is also flexibility in the design of the Protocol (it is not a treaty), as it allows for amendment by written agreement of the parties in Article 11 (US and Canada 2012). The ambiguous language of the Protocol, using such terms as 'when appropriate' and 'as appropriate' (Jetoo and Krantzberg 2014) leaves room for discretion of the implementing agencies in deciding on action most relevant to their local scale in achieving the measures stipulated in the Protocol. However, this can also be seen as a weakness, as there are no sanctions for not achieving the measures in the Protocol and this can retard progress. The vague language can also be seen as a hindrance to the role of the IJC, as it does not have a standard against which to compare 'as appropriate' measures by the parties in meeting their obligations under the Protocol.

There are many parallels for HELCOM, with the Commission conducting research to lend acceptance of new ideas. It was this research that lead to the realization of the Baltic Sea Action Plan. Monitoring information that is used by HELCOM for its assessments and reports are done by member countries and by the Monitoring and Assessment sub-group (MONAS) (HELCOM 2013). There is provision for revision of the BSAP at the ministerial meetings, lending it flexibility. Although BSAP recommends specific nutrient reduction targets for members, there is no stipulation as to how these goals are to be met. This lends flexibility to the means of achieving the programs, which leads to greater adaptability. However, such flexibility makes it difficult to assess progress towards the goals and may not be stringent enough to motivate states to implement nutrient reduction programs. This is achieved, however, to a high extent through EU environmental and marine legislative instruments as this has been introduced over time with a more ecosystem approach management to the governance of EU environment (Söderström and Kern 2017). There is no mechanism for regular evaluation of member's performance by HELCOM as is done by the IJC and no mechanisms for binational or multination cooperation, as is the spirit of the IJC. Whilst HELCOM is founded by the binding international law agreement, the Helsinki convention, the ambiguous language gives member states great flexibility in the implementation of measures. There is ambiguity in measure to be taken and what are deemed appropriate measures.

Resources

In order to fulfil their mandates, both the IJC and HELCOM require financial, human and technical resources. Financial resources is needed both for facilitation of meetings and for conducting individual studies on pressing issues, if the need arises. Article 4 (clause 3b) of the Protocol makes provision for funding for the IJC stating that the parties commit themselves to seek "the appropriation of funds carried out by the International Joint Commission to carry out its responsibilities effectively" (US and Canada 2012), with each party paying one half of the budget for the Great Lakes Regional Office. Expenditures for the Canadian section of the IJC (will be an equivalent amount for the US section) were c$ 6,761,044, with c$ 6,772,067 as the projected budgetary estimate for 2016–2017 (Canada 2016). This budgetary amount is for funding for the IJC's work under the Great Lakes Water Quality Agreement and for other transboundary water projects. For example, it includes funding for water quality modelling in the Great Lakes Basin and for assessment of residential flooding potential under the 1938 Rainy Lake Convention. Funding seems to be flowing smoothly for the operations of the IJC, as US$ 50K was recently approved for the preparation of a science synthesis on the impacts of petroleum transportation on the Great Lakes water quality (IJC 2016). Although this is a positive trend, the history of allocation of resources by the parties has shown that funding can be slow and uncertain (Commissioner for Environment and Sustainable Development 2001). Reduction of funding to the IJC since the 1990s in the past has led to a loss of the reporting role of the IJC (Botts and Muldoon 2005). For the IJC to be effective in executing its mandate under the Protocol, there needs to be adequate financing as the commission is more effective with more financing (Jetoo and Krantzberg 2015).

HELCOM is funded through the members' contribution to the convention and by other EU funding. According to HELCOM (2017b), a special fund has been set up to help with implementation of the BSAP, which provides grants for technical assistance of projects such as wastewater treatment, nutrient recycling, manure management and reduction of environmental pollution from shipping. It is managed by the Nordic Investment Bank (NIB) and the Nordic Environment Finance Corporation (NEFCO), and disposes approximately 11 million euros, donated by the governments of Finland and Sweden. HELCOM can also access funding through EU initiative such as the EU strategy for the Baltic Sea region. There has been no record to date of inadequate access to funding hampering HELCOM's ability to perform its work, as in the case of the IJC. Whilst funding is readily available for HELCOM's activities, funding for implementation of measures by member countries may be hampering the effective implementation of the BSAP and thus, hampering the adaptive capacity of HELCOM to respond to change.

Discussion and Final Conclusions

Looking back to the determinants for adaptive capacity in our cases the authors can see that despite a common framework and to a high extent similar problem picture, there are variations to be found in our cases. These variations do highlight—to some extent—the geo-political situation in the respective region, as well as the historic foundations for the institutions.

Table 6.2 highlight the main findings for the determinants.

As can be seen, there are some evident differences in the functions of both institutions. The formal public participation mechanisms, that were present for IJC but weakened, are not formally present at all in HELCOM. This does highlight the basic difference in the institutions from start, HELCOM highlighting the multilateral state centric view from mid-1970s. On the other hand, IJC did highlight very early on the problem definition as something common, not only for US and for Canada, but also for the people in the region. Differences are today rather small, but still visible. It is clear that it is very difficult to provide members of the public opportunities to contribute meaningfully, but this could be enhanced through active, targetted outreach and webinars to more effectively engage the public.

The scientific foundation for both institutions seems to be solid, however leaving space for interdisciplinary knowledge as also local knowledge, especially in HELCOM through participatory mechanisms for experiential knowledge of individuals living in the Baltic Sea region, whose livelihoods are at stake (Haila et al. 2008).

Regarding networks we do find evidence in both cases of well-functioning, broad network structures. These are introduced at different stages, however, and network members might see their contribution inefficient to some extent. Observer status in HELCOM does not give the same influence as direct membership in different sub-organization structures (as in IJC). HELCOM could give more credit to the work of observer network. Efficient networking has the potential to influence on member countries implementation measures, with HELCOM acting as

Table 6.2. Determinants of adaptive capacity for IJC and HELCOM.

Determinant	Case IJC	Case HELCOM
D1 Public Participation Representation	Public provided influence in IJC through specific forums for participation, the effectiveness of this is though limited.	No formal public participation mechanism other than through observer status for organized interests. Possibility to follow meetings through application, but this is rather limited.
D2 Science	Solid and broad science as a base.	Solid and broad science as a base, also through research funding mechanisms outside the scope of HELCOM.
D3 Networks	Broad and inclusive participation structures for public and organized interests, to some extent limited though over time. Through formal mechanisms also efficient participation.	Broad network structure, active participation for organized interests as observers, less proof of efficient participation.
D4 Leadership Experience Decision Making	Bilateral agreement between democratic states, showing also binationalism with rather high level of individual membership in the steering board (3 per country). High focus also on multi-level mechanisms with sub-federal participation.	Multilateral agreement between states that did not have a common value base – making structures solid but also inflexible. Further development has changed the working environment highlighting the leadership of EU and introducing multi-level mechanisms to limited degree.
D5 Flexibility	In-built flexibility in implementation measures (vague language), working feed-back loop for science to enhance flexibility.	In-built flexibility in achieving goals, follow-up and evaluation deficit, however, changes over time indicate a certain level of flexibility.
D6 Resources	Basically solid funding scheme, but variations over time, cutbacks in activities due to this.	Sufficient resources and solid funding, implementation resources varying, but supported through common funding instruments.

a glue that binds different organizations in the region to work together. More efficient networking would mean more effective communication, leading to flexible adaptive actions in times of change, in both cases.

The multi-level implementation of environmental and marine policies in EU as well as the EUSBSR has highlighted the EU leadership in Baltic Sea governance, therefore also in HELCOM. This does give good options for policy implementation even though lack of assessment of measures by HELCOM to meet the BSAP is a clear obstacle, as well as the geo-political position of Russia. The federal structure of US and Canada can also be an obstacle given that the problem definition or available measures would differ between different regions in the Great Lakes area.

References

Aswani, S., Christie, P., Muthiga, N.A., Mahon, R., Primavera, J.H., Cramer, L.A., Barbier, E.B., Granek, E.F., Kennedy, C.J., Wolanski, E. and Hacker, S. (2012). The way forward with ecosystem-based management in tropical contexts: Reconciling with existing management systems. *Marine Policy*, 36(1): 1–10.

Berkes, F. and Folke, C. (1998). Linking social and ecological systems for resilience and sustainability. Linking Social and Ecological Systems: Management Practices and Social Mechanisms for Building Resilience, 1(4).

Berkes, F., Folke, C. and Colding, J. (2000). Linking Social and Ecological Systems: Management Practices and Social Mechanisms for Building Resilience. Cambridge University Press.

Botts, L. and Muldoon, P. (2005). Evolution of the Great Lakes water quality agreement. Michigan State University Press.

Boundary Waters Treaty, Jan. I1, 1909, U.S.-Gr. Brit., 36 Stat. 2448.

Canada 2016. 2016–17 Estimates. The Government Expenditure Plan and Main Estimates. Accessed on January 12, 2017 at 12:28 hrs at: https://www.tbs-sct.gc.ca/hgw-cgf/finances/pgs-pdg/gepme-pdgbpd/20162017/me-bpd-eng.pdf.

Commissioner of Environment and Sustainable Development. (2001). The International Joint Commission: A Key Binational Organization. Report from the Office of the Auditor General of Canada.

Convention on Biological Diversity (CBD). (1995). CBD II/8 "The Second Meeting of the Conference of the Parties (COP) of the Convention on Biological Diversity. Decision 8." UNEP/CBD/COP/2/19, November 1995, at p. 12.

COP5. (2000). Fifth Meeting of the Conference of the Parties to the Convention on Biological Diversity, 15–26 May 2000, Nairobi, Kenya, Decision V/6.

Dreisziger, N.F. (1981). Dreams and disappointments. pp. 8–23. Spencer, R.A., Kirton, J.J. and Nossal, K.R. (eds.). The International Joint Commission Seventy Years On, Centre for International Studies, University of Toronto. Ca. Accessed on January 9, 2017 at 17:18 hrs at: https://www.cbd.int/decision/cop/default.shtml?id=7148.

Ehlers, P. (1994). The Baltic Sea area: convention on the protection of the marine environment of the Baltic Sea area (Helsinki Convention) of 1974 and the revised convention of 1992. *Marine Pollution Bulletin*, 29(6-12): 617–621.

Gordon, W. Brown. (1948). The growth of peaceful settlement between Canada and the United States (Toronto 1948). p. 26. *In*: Spencer, R.A., Kirton, J.J. and Nossal, K.R. (eds.) 1981. The International Joint Commission Seventy Years On. Centre for International Studies, University of Toronto. Ca.

HELCOM. (1974). Convention on the Protection of the Marine Environment of the Baltic Sea Area, 1974 (Helsinki Convention). Accessed online on January 9, 2017 at 13:36 hrs. at: http://helcom.fi/Documents/About%20us/Convention%20and%20commitments/Helsinki%20Convention/1974_Convention.pdf.

HELCOM. (2013). Rules of Procedure. Accessed on HELCOM website on Jan 12, 2017 at 16:25 hrs. at: http://helcom.fi/Documents/About%20us/Internal%20rules/Rules%20of%20Procedure%202013.pdf.

HELCOM. (2017a). Convention on the Protection of the Marine Environment of the Baltic Sea Area, 1992. Accessed on HELCOM website, on Jan 11, 2017 at: http://helcom.fi/about-us/convention.

HELCOM. (2017b). Baltic Sea Action Plan: Reaching Good Environmental Status for the Baltic Sea. Accessed on HELCOM website on Jan 12, 2017 at 14:12 hrs. at: http://helcom.fi/baltic-sea-action-plan.

HELCOM. (2017c). Observers. Accessed on HELCOM website on Jan 12, 2017 at 15:52 hrs. at: http://www.helcom.fi/about-us/observers/international-non-governmental-organisations/.

HELCOM. (2017d). News. Accessed on HELCOM website on Jan 12, 2017 at 17:29 hrs at: http://www.helcom.fi/news/Pages/More-transparency-and-political-leadership-called-for-in-HELCOM-seminar.aspx.

Haila, Y., Joas, M., Jahn, D. and Kern, K. (2008). Unity versus disunity of environmental governance in the Baltic Sea Region. Governing a common sea. *Environmental Politics in the Baltic Sea Region*. Earthscan, London, pp. 193–212.

Intergovernmental Panel for Climate Change (IPCC). (2007). Climate Change 2007: Working Group II-impacts, Action and Vulnerability. IPCC Fourth assessment report: Climate change 2007. Accessed on Jan 8th, 2017 at 18:00 hrs. at: http://www.ipcc.ch/publications_and_data/ar4/wg2/en/ch18s18-6.html.

International Joint Commission (IJC). (2016). Minutes of the International Joint Commission Executive Meeting Windsor, Ontario June 20–22, 2016. Accessed online on Jan 8, 2017 at 12:44 hrs. at: http://www.ijc.org/files/publications/2016_06_20-22_en.pdf.

Jetoo, S. and Krantzberg, G. (2014). A SWOT analysis of the Great Lakes water quality protocol 2012: The good, the bad and the opportunity. *Electronic Green Journal*, 1(37).

Jetoo, S and Krantzberg, G. (2015). The Great Lakes water quality protocol 2012: A focus on the effectiveness of the International Joint Commission. *The International Journal of Sustainability in Economic, Social and Cultural Context*, 2(11).

Jetoo, S. and Krantzberg, G. (2016). Adaptive capacity for eutrophication governance of the Laurentian Great Lakes. *Electronic Green Journal*, 1(39).

Olsson, P., Gunderson, L.H., Carpenter, S.R., Ryan, P., Lebel, L., Folke, C. and Holling, C.S. (2006). Shooting the rapids: navigating transitions to adaptive governance of social-ecological systems. *Ecology and Society*, 11(1): 18.

Pahl-Wostl, C. (2009). A conceptual framework for analysing adaptive capacity and multi-level learning processes in resource governance regimes. *Global Environmental Change*, 19(3): 354–365.

Pentland Ralph. (2015). Ten Year Review of the International Joint Commission's Report on Protection of the Waters of the Great Lakes. Accessed online on January 11, 2017 at: http://ijc.org/files/tinymce/uploaded/Consultants_Report_Ten_Year_Review_of_the_IJCs_Report_on_PWGL_December_2015.pdf.

Räsänen, T. and Laakkonen, S. (2008). Institutionalization of an international environmental policy regime: the Helsinki Convention, Finland and the Cold War. Governing a Common Sea. *Environmental Policies in the Baltic Sea Region*, pp. 43–59.

Schindler, D.W., Armstrong, F.A.J., Holmgren, S.K. and Brunskill, G.J. (1971). Eutrophication of Lake 227, Experimental Lakes Area, northwestern Ontario, by addition of phosphate and nitrate. *Journal of the Fisheries Board of Canada*, 28(11): 1763–1782.

Söderström, S. and Kern, K. (2017) (forthcoming). The Ecosystem Approach to Management in Marine Environmental Governance: Institutional Interplay in the Baltic Sea Region (submitted to Environmental Policy and Governance 5 of September 2016).

Szaro, Robert C., William, T. Sexton and Charles R. Malone. (1998). The emergence of ecosystem management as a tool for meeting people's needs and sustaining ecosystems. Landscape and Urban Planning 40(1): 1–7.

US and Canada. (1978). Great Lakes Water Quality Agreement of 1978. The International Joint Commission. US and Canada.

US and Canada. 2012. The Great Lakes Water Quality Agreement Protocol 2012. US EPA and Environment Canada Publication.

VanDeveer, S.D. 2011. Networked Baltic environmental cooperation. *Journal of Baltic Studies*, 42(1): 37–55.

Wuff, F., Sokolov, A. and Savchuk, O. 2013. Nest-a Decision Support System for Management of the Baltic Sea. Technical Report No. 10. Baltic Nest Institute.

CHAPTER 7

Learning from the Transboundary Governance of Lake Victoria's Fisheries

Ted J. Lawrence,[1,*] James M. Njiru,[2] Fiona Nunan,[3] Kevin O. Obiero,[2] Martin Van der Knaap[4] and Oliva C. Mkumbo[5]

Introduction

Like many freshwater systems around the world, Lake Victoria has changed dramatically since the start of the 20th century due to opportunities and challenges associated with open-access, competition for common-pool resources, environmental degradation (Njiru et al. 2008), introduction of exotic species (Njiru et al. 2005), eutrophication (Njiru et al. 2008), and overfishing (Mkumbo et al. 2007). Exacerbating the stresses on freshwater resources is climatic changes (Winfield et al. 2016) leading to high human migration rates toward this water resource as previous arable lands turn into desolate lands, forcing migration towards cities and lakes (WWAP

---

[1] African Center for Aquatic Research and Education & Great Lakes Fishery Commission, 2100 Commonwealth, Blvd. Suite 100, Ann Arbor, MI 48105.
[2] Kenya Marine and Fisheries Research Institute, Mombasa Research Center, P.O. Box 81651 – 80100, Mombasa, Kenya.
Email: jamnji@gmail.com; kevobiero@gmail.com
[3] International Development Department, School of Government and Society, University of Birmingham, Edgbaston, Birmingham, B15 2TT, United Kingdom.
Email: F.S.Nunan@bham.ac.uk
[4] Food and Agriculture Organization of the United Nations, Regional Office for Africa, PO Box 1628, Accra, Ghana.
Email: martin.vanderknaap@hotmail.com
[5] Lake Victoria Fisheries Organization, Plot 7B/7E, Busoga, Square/P.O.Box 1625, Elizabeth Rd, Jinja, Uganda.
Email: ocmkumbo@lvfo.org
* Corresponding author: ted@agl-acare.org

2012). The human population surrounding Lake Victoria, thus, is growing in many of the basin's cities as a result of erratic rains, poor soils, crop failures and high unemployment throughout the region (Odada et al. 2009). This migration threatens the integrity of Lake Victoria and other water-scarce areas of the world where populations are beginning to lose access to clean, freshwater resources, thus exacerbating already stressed resources, the ability to govern them, and highly unequal access to these resources.

Due to the importance of the lake's resources for millions of people living in the basin, efforts to address environmental issues have, and are, taking place at regional, national, and local levels. Two current, notable, transboundary efforts are currently taking place on Lake Victoria, with the most comprehensive, the focus of this chapter, directed towards managing Lake Victoria's fishery resources. The other transboundary effort, conducted by the Lake Victoria Basin Commission, casts a wider net and includes tourism, agriculture, industry, and transportation in its purview.

The fisheries governance and co-management systems on Lake Victoria were created to address challenges of over-harvesting due to increasing capacity and the use of illegal methods to harvest common pool resources, informed by theoretical concepts of governance, decentralization, sharing authority, community participation (Béné and Neiland 2006), and Socio-Ecological Systems (SESs) (Ostrom 2009). Thus the co-management system, whereby the central government shares responsibility, authority, and resources with the communities, was established to motivate and reinforce legal fishing behavior and, therefore, produce more sustainable fisheries and sustainable development (Bwathondi et al. 2001). The concepts on which Lake Victoria's fisheries co-management system were created include:

1. decentralization, an approach where authority is formally ceded to lower political or administrative levels from central government;
2. sharing authority, or power-sharing, a result of decentralization, where responsibility of managing the resource is devolved and shared between government and community;
3. community participation, an approach where local communities and other stakeholders become active in management of natural resources; and
4. self-organization—a concept of co-management where local actors act on behalf of the fisheries sector in cooperation with other entities (e.g., the government) or with little or no external assistance or influence if necessary.

The system was designed to sustainably manage the fishery resource while providing a mechanism for human and community development. The system is a decentralized, community-based, co-management approach which includes the three riparian countries of Lake Victoria (Uganda, Kenya, and Tanzania) and their departments of fisheries, fisheries research institutes, the private sector[1] and the communities themselves. All of these entities fall under the coordination of the Lake Victoria Fisheries Organization (LVFO). The Lake Victoria Fisheries Organization provides for the harmonization of rules between each of the three countries and assists with the dissemination and capacity building of fisheries management regulations and programs. The communities have been empowered through organizations called Beach Management Units (BMUs). The BMUs are community run, governing bodies which assist in managing the fishery. The co-management system is designed to transcend resource issues across national borders and to overcome the differences in fisheries management that come with varying degrees of decentralization between the lake's three governments. BMUs have been created to change the roles of the local fishers, and their behavior concerning the use of illegal gear in an effort to sustain the fish stocks and create a sustainable fishery.

Though efforts to address threats to the lake are broad, the comprehensive focus of addressing fisheries management on the lake gives an in-depth example of transboundary cooperation. In this chapter, therefore, the authors discuss influences which have limited past management attempts of Lake Victoria's fisheries resources, the impetus for the current management approaches, the current management system, and ongoing challenges that need to be addressed.

Lake Victoria and the Fisheries Resources

Lake Victoria is the largest lake by surface area in Africa, and, like most freshwater resources, it is an important multi-use resource, known distinctly for its valuable, vibrant, and diverse fisheries. Lake Victoria provides employment for three million people in fisheries-related activities and is known for producing annual catches of around one million tons, contributing USD 840 million annually to the economies of Uganda, Kenya, and Tanzania (LVFO 2015a, Njiru et al. 2008), and contributes to food security and poverty reduction to 20 million people around the lake (LVFO 2011a, Ugandan Department of Fisheries Resources 2003). Due to its tropical setting, Lake Victoria's capture fisheries produce more fish than

---

[1] Representatives of associations of fish processing plants are involved in the process.

the commercial fisheries on all five Laurentian Great Lakes combined, five times the harvest of Lake Tanganyika (LTA 2012), more than quadruple the harvest of Lake Malawi/Nyasa/Niassa (FAO 2012), the second and third largest lakes in Africa, respectively.[2]

Fishing pressure on Lake Victoria is high, with an estimated 1,530 fish-landing-sites, 206,425 fishers, and 70,399 fishing craft in 2014 (LVFO 2015b). Fishing pressure is, in part, a result of a lack of viable alternative livelihoods; the demand for fish protein; increasing value of fresh and processed fish globally; increasingly efficient fishing technology (Mkumbo et al. 2007), demonstrated by the mechanization of the fishing fleet which has increased from 4,108 in 2000 to a total of 21,578 motorized fishing units in 2014; and, the use of various types of illegal gear which prevent regeneration of stocks. The human population surrounding the lake, and hence pressure on the region's environmental and fisheries resources, is growing at 3% per annum (Odada et al. 2009).

To stay abreast of changing ecological, social, and political factors, fisheries management on Lake Victoria has had to evolve from parochial and disparate attempts at fisheries management to a collaboration between riparian states and communities, recognizing the essential role and stake that communities who rely on these resources have in its management. The changes, therefore, that influenced the creation of Lake Victoria's current fisheries management system, and became main pillars of the management system, included the shift to a transboundary approach with the intent of harmonization and the adoption of a participatory approach to the governance and management of these resources which is deemed necessary based on attributes of the fishers and the natural resources (Gaden et al. 2012). The shift towards a more participatory and decentralized fisheries management in the late 1990s followed evidence that a centralized, top-down model of fisheries management had not stemmed declines in fish stocks (Etiegni et al. 2016).

A core challenge in managing Lake Victoria is that its natural resources are embedded in complex, social-ecological systems (SESs). As a valuable resource shared by three East African States, Lake Victoria has long-standing and complex governance arrangements involving numerous government agencies and a variety of users and stakeholders that have evolved over

---

[2] Compared catch data from Great Lakes total commercial fisheries catch (45,454 tons/year) source: (http://www.newton.dep.anl.gov/natbltn/200-299/nb295.htm) with 1999 catch data from Lake Victoria 700,000 to 1,000,000 t of fish (FAO 2010a–e, LVFO 2012); the Great Lakes provides about 118,430 jobs in agriculture, fishing and food production (Vaccaro and Read 2011). Lake Tanganyika's total annual harvest is between 110,000–120,000 tons in 2012 (Van der Knaap in press); Lake Malawi's total annual harvest is between 40,000–50,000 tons/year (assuming Lake Malawi produces 75 percent of total fisheries catch in Malawi) years 2000–2009 (FAO 2012).

the past 100 years. In a complex SES, multiple subsystems such as the resource system (e.g., the fisheries or water resources), resource units (e.g., fish), users (fishers), and governance systems (organizations and rules that govern fishing on the lake) interact to produce complex outcomes, which in turn require knowledge about how these specific subsystems and their component parts are related (Ostrom 2009). Therefore, to protect the fisheries, livelihoods, and well-being of those who depend upon them, formal fisheries management system was developed to include fishers and other community members engaged in fisheries (Lawrence 2015).

Fisheries Management on Lake Victoria, East Africa

The Impetus for Current Fisheries Management System on Lake Victoria

From 1885 through the 1950s, the influence of emerging markets and extractive policies of the colonial governments of East Africa, the introduction of more efficient fish capture technology (larger boats, nets), access and transport to the lake's resources (rail and roads), and innovation (engines) combined with increasing human populations around the lake, led to the need for social and political institutions to effectively manage the fisheries. Fisheries management of the lake dates back to 1908, when the Fish Protection Ordinance to ban specific gear was enacted (Geheb 1997). In 1947, the Lake Victoria Fisheries Service was formed to enforce fisheries laws and regulations (LVFO 2001). This was later followed by the East Africa Freshwater Fisheries Research Organization (EAFFRO) in 1960, which was disbanded when the East African Community (EAC) was dissolved in 1977. The first fisheries management efforts on Lake Victoria ended with the termination of the Lake Victoria Fisheries Service in 1960, with the roles of this institution transferred to the national fisheries departments of each partner state; EAFFRO's collapse emerged into individual research institutes (Ogutu-Ohwayo 2001) which currently exist today as the National Fisheries Resources Research Institute (NaFIRRI) in Uganda, Kenya Marine and Fisheries Research Institute (KMFRI), and Tanzania Fisheries Research Institute (TAFIRI). Following the breakup of the first EAC, the Food and Agriculture Organization of the United Nations (FAO) coordinated research and management efforts on Lake Victoria through its Committee on Inland Fisheries of Africa (CIFA), especially the Sub-committee for the Management of Lake Victoria, which in 1994 led to the establishment of the Lake Victoria Fisheries Organization (LVFO). The LVFO was established to harmonize and coordinate the sustainable management of the fisheries resources of Lake Victoria.

Prior to the 1994 LVFO Convention, the management approach on Lake Victoria was based on top-down enforcement of management measures, with little consultation with, or participation of, fishing communities (Van der Knaap et al. 2002). During colonial intervention on Lake Victoria, beginning in the 1890s, research on the fisheries was conducted by the colonial governments to determine the effects of increased fishing pressure on the resource and colonial-led management programs attempted to control excessive harvest of the lake's resources.

During the late 1950s and early 1960s, toward the end of colonial authority in East Africa, the colonial government of Uganda introduced the exotic, piscivorous, species Nile perch (*Lates niloticus*) in Lake Victoria. The decision to stock this fish was driven by the colonial administrators who favored converting the native Haplochromine cichlids[3] (still thriving in the 1950s) into something suitable for the European restaurant table and making the Lake Victoria fishery into an economically powerful industry, especially with prospects of a highly valued European export market (Mkumbo and Mlaponi 2007). The haplochromine fish, widely depended upon by the local fishermen, were considered bony "trash" fish by the Europeans (Anderson 1961).

The emergence of the Nile perch increased fish catches tremendously, transforming Lake Victoria's fishery from artisanal to commercial, increasing fishers' income and employment opportunities (Njiru et al. 2005). Regardless of its perceived success as a commercial species, the introduction of Nile perch was opposed by fishery ecologists, as they were apprehensive about the unknown impacts that a top-predator would have on Lake Victoria's complex ecosystem (Fryer 1960, Kudhongania et al. 1992, Pringle 2005). The ecological changes that occurred in Lake Victoria are still being studied today and continue to change because of social, political, and economic variables regarding the continued importance (economically) of the Nile perch for the nations bordering Lake Victoria and the fishers who rely on them for their livelihood.

Though numerous factors such as overfishing and pollution have impacted fish populations on Lake Victoria, the decline of Nile perch populations was the impetus for reinvigorated management efforts on Lake Victoria in the 1990s (Lawrence 2015). The demand for Nile perch resulted in increased fishing efforts, overfishing, and illegal fishing that reduced Nile perch populations (Mkumbo and Mlaponi 2007, Mkumbo et al. 2007, Njiru et al. 2007, Ojuok et al. 2007, Witte et al. 2007). The decline of these valuable stocks necessitated the riparian governments to include communities in what is referred to as co-management. Fishery co-management efforts were a result, in part, of past failed centralized

---

[3] A species-rich group of small, perch-like fish of the family *Cichlidae*, genus *Haplochromis* (Goldschmidt 1998; p. 3). Lake Victoria's fish fauna was dominated by over 500 species of *Haplochromis cichlids* prior to 1980 (Witte et al. 2007).

federal government efforts on Lake Victoria and, in general, a shift by governments in developing countries to move toward community-based management of natural resources (Lawrence 2015, Obiero et al. 2015).

Between the time of colonial intervention and the 1990s, when the Nile perch populations were large enough that commercial exploitation could take place, several fisheries management programs or sets of fisheries regulations were attempted. Among them were the mesh size and license regulations as described by Dobbs (1927). In subsequent decades certain fishing gears and methods were banned; minimum mesh sizes were adjusted with respect to changing species compositions in the catches. With the start of the commercial fisheries for Nile perch, combinations of measures were proposed by the Haplochromis Ecology Survey Team (HEST) (1989) and CIFA (1994), ranging from mesh sizes in gillnets, to closed areas and seasons. Eventually, the aim was to reduce fishing effort. Subsequently a slot size for the perch was introduced, from 50 to 85 cm total length, but only the lower limit was self-enforced by the processing and exporting industry. In addition to measures to restrict fish harvests, programs were established aimed at strengthening research institutions and bolster research on Lake Victoria. Examples include the European Union-funded Lake Victoria Fisheries Research Project (LVFRP), aimed at strengthening the research institutions and research, and the Lake Victoria Environmental Management Project (LVEMP), and the Implementation of a Fisheries Management Plan (IFMP), which worked towards putting in place and strengthening the institutional structures and management systems that currently exist. Many of the past measures that were established to limit fish harvest, using gear size or type restrictions, slot sizes, and other constraints, were irregularly enforced, apart from a trawl ban, which was eventually enforced lake wide (Van der Knaap et al. 2002).

The introduction of the invasive Nile perch in the 1950s drastically changed management approaches on Lake Victoria due to its prolific reproductive capacity and subsequent high value (Pringle 2005). Between 1960 and 1981, the biomass of Lake Victoria changed from 80% native haplochromine species to 80% Nile perch (Kudhongania and Cordone 1974, Ogutu-Ohwayo 1990, Witte et al. 2007). By the mid-1980s, increased fishing by the rapidly growing human population necessitated a new fisheries management approach as harvest of the fisheries resources was conducted in a manner unsustainable for future generations. Van der Knaap and Ligtvoet (2010) considered the exploitation of Nile perch on Lake Victoria sustainable in the first years of the current millennium, but subsequently the fishing capacity targeting the Nile perch increased rapidly, not only the numbers of fishermen and canoes, but also the sizes of the gillnets and the numbers of longline hooks. As the commercial prospects of Nile perch were being over-fished, further efforts to manage the resources were established.

The Lake Victoria Fisheries Organization: An Interjurisdictional Organization

The Lake Victoria Fisheries Organization was formally established by Convention that entered into force on 24 May 1996 between the riparian states which share Lake Victoria: Uganda, Kenya, and Tanzania.[4] The Convention is a legal document with instruments of accession submitted by the three EAC riparian countries and established the LVFO as an international regional organization to advance cooperation between them; its aim is to "harmonize national measures for the sustainable utilization of the living resources of the Lake and to develop and adopt conservation and management measures" (LVFO Convention, Article II (2), 2001). The LVFO Convention is accommodated under *Article 9(3)* of the EAC Treaty. In 2016, the convention was amended by the EAC partner countries to expand the LVFO's mandate and scope from Lake Victoria (Uganda, Kenya, and Tanzania) to cover all fisheries and aquaculture within the EAC (Uganda, Kenya, Tanzania, Rwanda, and Burundi). The LVFO, thus, is a specialized institute of the EAC and serves the important function of coordinating and harmonizing fisheries management and research. The LVFO's function serves to address challenges that common-pool, multijurisdictional and transboundary fisheries present.

The LVFO secretariat is the executive structure which facilitates the processes for policies and decisions created at the highest levels of the LVFO—Council of Ministers[5] (LVFO 2001, 2011b). The LVFO also provides a forum to assess the implementation of fisheries management at the local level. The fishing regulations are mostly directed at the local user—fishers—who make the ultimate decision about how they will harvest the resource; rules exist to deter the fishing of undersized fish, including a ban on the trade in small Nile perch. The LVFO—through the national departments of fisheries—are charged with educating fishing communities (BMU committees and members of BMUs around the lake) about rules and how those rules aim to achieve sustainable fisheries and community development, the functions and responsibilities of BMUs and of fishers, and other functions and responsibilities of the fishing communities.

---

[4] At this time the EAC consisted of Uganda, Kenya, and Tanzania. Burundi and Rwanda were acceded into the EAC in 2007 and joined the LVFO in 2016; South Sudan was acceded into the EAC in 2016.

[5] The Council of Ministers (Council), consist of ministers of the partner states' ministries responsible for fisheries. The Council is presented recommendations by the Policy Steering Committee which is informed by a Fisheries Management Committee and Scientific Committee. Members of all committees are engaged in fisheries, including heads of fisheries departments, research institutions, and permanent secretaries responsible for fisheries (LVFO 2001, LVFO 2011b).

The governance structure of the LVFO (Fig. 7.1) includes the highest policy organ—the East African Community (EAC) Summit—to the grass root structures—the Beach Management Units (BMUs). The organizations that make up the LVFO, and are a part of the fisheries management system, are the national fisheries departments, research institutions, and the fishing communities that exist around the lake (Gaden et al. 2012). The

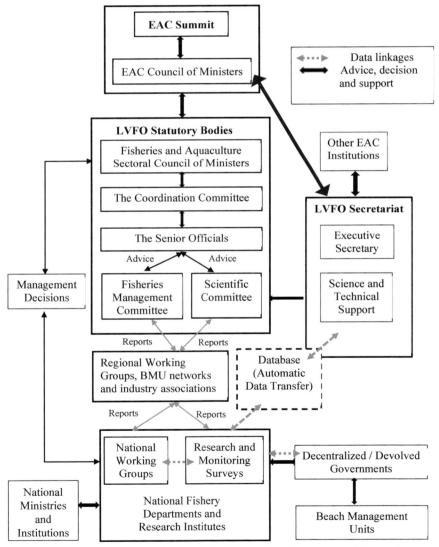

Figure 7.1. Organogram presenting the LVFO Institutional Structure and the Linkages to the Different Policy Organs at the Regional Level (Adapted from LVFO 2015a by K. Obiero).

private sector organized into Fish Processors and Exporters Associations at national and regional level are also involved in the management and participate in technical committees of LVFO as members.

The LVFO achieves regional coordination through two technical committees (LVFO 2001), the Fisheries Management and Scientific Committees, composed of the directors of fisheries departments and research institutions respectively; both committees form the Executive Committee. These committees serve as the technical arms of the LVFO, and are advised by fisheries management and research working groups (Gaden et al. 2012). The working groups conduct research and socio-economic studies related to the status of the stocks and the welfare of the fishing communities; this knowledge is then transferred to the LVFO's Executive Committee via the Fisheries Management or Scientific Committee. This Executive Committee in turn considers and recommends management measures to the "Policy Steering Committee, comprising [permanent secretaries] of the ministries dealing with fisheries matters in each of the contracting parties" (Gaden et al. 2012). Finally, the Policy Steering Committee submits recommendations to the Council of Ministers (ministers responsible for fisheries matters in the contracting parties), which adopts management and conservation measures. The LVFO committees hold one session annually to review progress and compliance in implementing the agreed measures. The measures are also revised as needed (LVFO 2001).

At the regional level, the LVFO plays the role of harmonizing measures, advocating for the resources and the users, and disseminating the agreed measures to the contracting parties after adoption by the Council of Ministers. The convention establishing the LVFO demands the contracting parties provide the organization with such laws, regulations and any other related documents and information for the purpose of assessing compliance or any disparities which will warrant adjustments and harmonization (Gaden et al. 2012).

Central Government: Ministries and Departments

Fisheries-related central authority on Lake Victoria are the Ministries of Agriculture, Animal Industry and Fisheries (Uganda), Ministry of Agriculture, Livestock and Fisheries (Kenya), and Ministry of Agriculture, Livestock and Fisheries Development (Tanzania). These ministries oversee their respective fisheries directorates[6] (DOF). Each central government around Lake Victoria—under trust law—is obligated to hold natural resources in trust for the people of their countries (Republic of Kenya

---

[6] Ugandan Directorate of Fisheries Resources, Tanzanian Fisheries Development Division, Kenyan State Department of Fisheries and Blue Economy.

2008, Ugandan Ministry of Agriculture Animal Industry and Fisheries 2004, United Republic of Tanzania 1977). Legal ownership of fisheries resources is vested in the states' central government authority as trustee and, therefore, the state is obliged to manage the resource in the interest of beneficiaries, those who depend on these resources (Naluwairo 2005).

Guidelines for fisheries management on Lake Victoria state that each partner state's department of fisheries promote, support, and guide BMUs ability to function, especially under "circumstances where local capacity alone will not be sufficient to safeguard the livelihoods of people depending on fisheries resources" (Ugandan Ministry of Agriculture Animal Industry and Fisheries 2004). However, as is noted throughout this chapter, support by government of community entities can be weak or non-existent.

Mid-level Entities

For the purpose of this chapter, mid-level entities are referred to as the Local Government authorities in each country and any entity that join the Fisheries Officers and BMUs for monitoring, control, and surveillance (MCS) purposes. Fisheries officers are located at the decentralized levels of governance, including district (Tanzania and Uganda) and County level (Kenya) and below. The majority of fisheries officers are employed by the local government rather than the fisheries departments and so answer to the local government as well as the fisheries departments. They form an essential component of the co-management system, overseeing a region or district of BMUs and assisting in the enforcement of fishing rules, either in cooperation with, or for, BMUs. The officers engage in local government planning and budgeting, seeking support for their work in fisheries, including raising funds for MCS. The County system of government is relatively recent in Kenya, following the County Governments Act of 2012, and so the relative share of responsibility and revenue associated with fisheries is still in the process of being finalized. As the majority of fisheries staff is employed by the local government, this can bring challenges to the fisheries departments in terms of communication and influence over what fisheries officers do. In addition, there can be challenges to fisheries management in terms of interference by local politicians, particularly in terms of enforcing regulations during election periods (Nunan et al. 2016).

Other mid-level entities include police officers who often assist in fishing patrols, with the intent of ensuring the safety of community patrol members and fisheries staff and reinforcing the legitimacy of the BMU committee, and the judiciary, charged with enforcing legislation. The capacity for enforcement, however, is limited and officers from all these parts of government have been accused of engaging in corruption associated with fisheries illegalities (Lawrence 2015, Nunan et al. 2016).

Local Fisheries Management: Beach Management Units

Beach Management Units were created during a global shift to involve communities in the management of their own natural resources in developing countries, switching from top-down approaches of governance—whereby government agencies managed the resources—to collaborative approaches—whereby communities are involved in making decisions concerning the management of the resources and thus gain an increased sense of ownership and management of compliance. On Lake Victoria, this approach coincided with the booming Nile perch industry and BMUs were formed from the late 1990s to help regulate fish processing activities at landing sites where Nile perch were landed. The facilities are meant to ensure the fish are hygienically treated for transport to fish processing plants for export to international markets. BMUs became the foundation of Lake Victoria's fisheries co-management system. The BMUs were created so that rules could be standardized and enforced around the shores of the lake, while reducing the constant exercise of the limited enforcement by the government, and internalize patterns of behavior with roles, rules, regulations, and goals that are clearly defined and created by the communities. The BMUs are legally empowered, local community organizations that are mandated to: enforce government fisheries regulations; create and enforce their own local by-laws—governed by lake-wide guidelines—for sustainable fisheries management; serve as resource-data collection points for better fisheries management and monitoring; issue fish transport licenses (revenues from which are used for fisheries management activities and may, in part, be retained by the BMUs for operations); increase local users' capacity to manage their finances; and resolve conflicts among BMU members, such as disputes over gears, and among fishermen and traders, boat owners and fishermen (Ebong et al. 2004, Obiero et al. 2015).

Each BMU is located around pre-existing, local fish-landing sites, though may consist of more than one landing site (LVFO 2005a). The BMUs are community-run governing organizations consisting of all users engaged in fisheries-related activities at a landing site, including "boat owners, fishing crew members, fish mongers, artisanal fish processors, local gear makers and repairers, boat builders, fishing input suppliers, and industrial fish processors' agents" (LVFO 2005a). Following notions of democratic decentralization, the fishing community must elect a BMU committee of 9–15 members to be drawn from their population and be inclusive of representatives of all stakeholder groups. Specifically, guidelines of membership of the committee specify that representation on the committee should follow a distribution of 30 percent boat owners; 30 percent crew (fishing laborers); 30 percent other (including fish processors, boat makers, local gear makers or repairers, and fish equipment sellers); and 10 percent fish mongers and traders (LVFO 2005a). Measures are therefore in place to include those who were historically underrepresented or disenfranchised, including women, boat crew, fish mongers, and traders (LVFO 2005a, 2011c). In addition, BMU committee guidelines require 30 percent membership of women, so as to empower women and consider their views in management decisions, especially based on their influence in the movement of fish and fisheries products (Kenyan Department of Fisheries 2006, Ugandan Department of Fisheries Resources 2003, United Republic of Tanzania 2011).

Each BMU on Lake Victoria is bound by similar fishing rules and guidelines, informed by all partners within the system. Country-specific fishery management differences exist (e.g., closed season for dagaa in Kenya). The BMU operates within pre-determined geographic boundaries, and assists with policy development, rules enforcement, and administrative duties pertaining to fisheries-related activities. BMU committees and networks are found at the local, regional, and national levels, thereby allowing BMU representation at higher levels of governance (Nunan 2010).

Fisheries management, such as enforcement of rules, patrolling for illegal gear and activities, and tax collection, are executed by the BMU committees. When done appropriately, these activities are carried out in "collaboration with the relevant authorities" (LVFO 2005a). It is at the local level, the interface between the BMU committee (regulators) and fishers, that fisheries harvest is regulated and the resultant tax collection is conducted.

On enforcement of fisheries rules, BMUs' major charge is specifically to "ensure compliance with local and national regulations . . . formulate and enforce community by-laws at the local level; [and] monitor fishing activities within their localities" (LVFO 2005a). BMUs therefore work collaboratively with government authorities (departments of fisheries) and enforcement institutions (police) to conduct patrols. Police and Fisheries Officers are involved for the security of community members and so that BMU members can arrest, and then prosecute offenders.

Discussion

Co-management and Transboundary Approaches: Lake Victoria and the Basin

Many motivations influenced the creation of current fisheries co-management on Lake Victoria. The process has occurred during the past few decades and has faced considerable challenges, including overfishing and high levels of non-compliance of regulations; ineffective top-down governance of fisheries management approaches; increasing human migration to freshwater resources and subsequent increased fishing harvest effort; and increasing global value of fisheries, especially that of the internationally sought after Nile perch.

Acknowledging that fisheries harvest by one riparian country has impacts on the other partner states (LVFO 2005b), the LVFO was created to harmonize fisheries resource approaches across national borders and to overcome differences in fisheries management resulting from varying, and disparate degrees of fisheries management between the lake's three national governments. Additionally, because of the reliance of a high number of community members on the fisheries, fishers and stakeholders engaged in fisheries activities were included in the management program, thus initiating co-management. The LVFO harmonized the existing formal constraints and incentives established by the lake's riparian countries and augmented them in the form of regulations intended to structure human interactions and control fishing harvest behavior to ensure sustainable fisheries and community and country-wide development (Geheb and Crean 2003, North 1990), allowing local communities to help manage their shared fisheries.

As described in this chapter, the fisheries of Lake Victoria are important to millions of people in, and beyond, the basin. The various uses of Lake Victoria, however, means there are also a diverse set of impacts on the lake and the fisheries (Nunan and Onyango 2017). Regardless of the specificity of focusing on the fishery, increased human populations around Lake Victoria have contributed to multiple environmental stressors (Hecky et al. 2010) which impact all of the lake's resources, including water quality for drinking and access to the lake's edge, environmental degradation of wetlands and other spawning areas (Njiru et al. 2008), introduction of exotic species (Njiru et al. 2005), eutrophication (Njiru et al. 2008), water levels from hydropower use (Lubovich 2009), fishery issues relating to transboundary conflict (Nunan and Onyango 2017), and aquaculture issues (Wedig and Stoehr in press). All of these issues have the potential to negatively influence the fisheries and have been acknowledged by fishers, scientists, and other stakeholders in Lake Victoria's fisheries system (LVFO 2005b).

Although the LVFO is fisheries-oriented, it is influential on other Lake Victoria-related issues, for example, it addresses transboundary issues through its Statutory Organs when they meet. During meetings the LVFO statutory organs discuss any emerging issues of transboundary nature. Once a conflict is officially reported, the LVFO provides a forum to discuss and resolve the matters. The LVFO, however, defers to the Local Government Authorities of both parties to first handle the issues.

Hydropower development is another example of water use that often does not focus on the importance of inland fish and fisheries and the peoples they support; human costs of hydropower development are rarely considered (Barlow 2016). On Lake Victoria, the LVFO has created water release policies which govern where, when and how much water can be abstracted from the Owen Falls Dam at the Source of the Nile, in Jinja, Uganda.[7] Regarding this structure, the Ugandan fisheries research institute (NaFIRRI), is assigned to consider the effect on the fisheries as a component of an Environment Impact Assessment. Also, on a regular basis, NaFIRRI monitors the fish stocks downstream and upstream of this dam.

One of the proposed steps for responsible inland fisheries is increased freshwater fish production through the sustainable intensification of aquaculture as referred to in the Rome Declaration: Ten steps to responsible inland fisheries (Taylor et al. 2016). Aquaculture is the fastest-growing food production sector and an important component in many poverty alleviation and food security programs. It can complement capture fisheries through stocking programs, by providing alternative livelihoods for fishers leaving the capture fisheries sector, and by providing alternative sources of animal protein. Aquaculture can also negatively affect capture fisheries through escape of invasive species and diseases, through competition for water resources, pollution, and access restrictions to traditional fishing grounds. Regarding aquaculture, LVFO has worked on comprehensive cage culture guidelines that are in use. In the Kenyan portion of Lake Victoria, approximately 6345 cages have been installed during the last year producing approximately 40,000 metric tons of fish LVEMP 2016. These guidelines are aimed at reducing some already observed negative aspects of aquaculture, such as cages that were installed

---

[7] Owen Falls Dam is the first dam and electric producing facility on the White Nile.

in shallow, breeding areas (disregarding scientific advice) and fish kills caused mainly by hypoxic conditions from aquaculture operations.

Other Transboundary Efforts to Manage Lake Victoria's Water Resources

Lake Victoria's water quality, like other freshwater lakes globally, depends on activities that take place outside of the lake itself. Agriculture, paper and sugar factories, mining, urban effluent, and other activities can discharge various pollutants into the lake, negatively affecting the quality of the water for human consumption and negatively affecting other aquatic resources, such as fish. As stated at the beginning of this chapter, there is another major effort taking place on Lake Victoria to address these broader transboundary water and environmental issues. The EAC's Lake Victoria Basin Commission (LVBC) focuses on more than just water issues, including tourism, agriculture, industry, and transportation. The EAC, indeed, stated that Lake Victoria and its basin be designated "as an 'area of common economic interest' and 'regional economic growth zone' to be developed jointly by the Partner States" (LVBC 2017).

To coordinate the various environmental health interventions on the lake and within the basin, the EAC established the LVBC in 2001 as a center to promote investments and information sharing for economic growth among the partner states. The LVBC's activities focus on: harmonizing policies and laws that pertain to the management of the lake's environment within its catchment; control and eradication of the invasive water hyacinth; the development of fishing, agriculture, and tourism industries; and, development of infrastructure including vitalizing transport system.

The LVBC harmonizes efforts between the five riparian countries of Lake Victoria. Through its Secretariat, the LVBC is concentrating more on the coordination and implementation of basin-wide projects meant to improve collaborative management of the transboundary natural resources of the Lake Victoria Basin among the partner states and to improve environmental management of targeted pollution hotspots and selected degraded sub-catchments for the benefit of communities who depend on the natural resources of Lake Victoria Basin (LVEMP-II). The LVBC is also involved, through United Nations Human Settlements Program (UN-HABITAT), in implementing a major initiative, the Lake Victoria Region Water and Sanitation Initiative (LV-WATSAN), to address the water and sanitation needs of the riparian population, particularly the poor, in the secondary urban centers around Lake Victoria.

Conclusion

Lake Victoria provides water, power, a transportation corridor, food, and employment for the people of the riparian countries that border it. The lake's fisheries provide a substantial protein source for communities in East Africa's riparian countries and for communities that extend well past these borders through international markets. The lake's management is essential for the health and benefit of the people living along its shores, within its basin, and beyond. The coordinated management of fisheries of Lake Victoria provides a good example of regional cooperation to overcome many of the management challenges on a transboundary resource of this size. Managing fish requires a coordinated, harmonized effort between multiple formal political jurisdictions, thousands of formal and informal communities, and multiple sectors. The complexity persists with the nature of the resource (invisible and mobile), the variable nature of the need for fish as a subsistence resource and an internationally, highly valued, sought after commodity.

Furthermore, the lake's resources depend on the health of the lake itself, and multiple sectors have varying impacts, including agriculture, urban development, forestry, and hydropower. The value of the lake's resources also can create conflict within and beyond fisheries, found in border areas, between fisheries-focused BMUs and villages, and among other stakeholders both within and without the fisheries sector.

The LVFO provides legal and social mechanism for government and communities to collaboratively manage the lake's resources on which millions depend. The management system on Lake Victoria has had many positive benefits for many of Lake Victoria's fishing communities, national development, and to some extent, reducing over harvest (Lawrence 2015, Njiru et al. 2006, Njiru et al. 2007, Njiru et al. 2014, Njiru et al. 2005, Nunan et al. 2016, Nunan et al. 2015, Nunan et al. 2012, Nunan and Onyango 2017, Obiero et al. 2015).

To do away with the system will likely result in tragedy-of-the-commons-type resource harvest behavior, whereby fishers would act by "the law of the jungle, anarchy, chaos, piracy, and conflict", the "fisheries would be bad and negatively affect the community" (Lawrence 2015). At the broader level, co-management should be seen as a process (Nunan et al. 2016), by which members of these institutions are given the capacity to adapt to ongoing changes (adaptive governance or adaptive co-management) and learn from experience elsewhere.

References

Anderson, A.M. (1961). Further observations concerning the proposed introduction of Nile perch into Lake Victoria. *East African Agricultural and Forestry Journal*, 26: 195–201.

Barlow, C. (2016). Conflicting agendas in the Mekong River: Mainstream hydropower development and sustainable fisheries. pp. 281–287. *In*: Taylor, W.W., Bartley, D.M., Goddard, C.I., Leonard, N.J. and Welcomme, R. (eds.). Freshwater, Fish and the Future: Proceedings of the Global Cross-sectoral Conference, Food and Agriculture Organization of the United Nations, Rome; Michigan State University, East Lansing; American Fisheries Society, Bethesda, Maryland.

Béné, C. and Neiland, A.E. (2006). From Participation to Governance: A critical review of the concepts of governance, co-management and participation, and their implementation in small-scale inland fisheries in developing countries. WorldFish Center Studies and Reviews 29. The WorldFish Center, Penang, Malaysia and the CGIAR Challenge Program on Water and Food, Colombo, Sri Lanka 72 p.

Bwathondi, P.O.J., Ogutu-Ohwayo, R. and Ogari, J. (2001). Lake Victoria Fisheries Management Plan (LVFMP). Socio-economic data working group of the Lake Victoria fisheries research project, Jinja. *In*: Cowx, I.G. and Crean, K. (eds.). Lake Victoria Fisheries Research Project (LVFRP) - Phase II: LVFRP Technical Document No. 16. LVFRP/TECH/01/16.

Dobbs, C.M. (1927). Fishing in the Kavirondo Gulf, Lake Victoria. *Journal of the East Africa and Uganda Natural History Society*, 30: 97–109.

Ebong, I., Lwanga, M. and Scullion, J. (2004). Beach management units and integrated lake management. *In*: Heck, S., Kirema-Mukasa, C.T., Nyandat, B. and Owino, J.P. (eds.). International Workshop on Community Participation in Fisheries Management on Lake Victoria. Kisumu, Kenya.

Etiegni, C.A., Irvine, K. and Kooy, M. (2016). Playing by whose rules? Community norms and fisheries rules in selected beaches within Lake Victoria (Kenya) co-management. *Environment, Development and Sustainability*, 1–19.

FAO. (2010a). Miscellaneous freshwater fisheries, capture production by species, fishing areas and countries or areas. Food and Agriculture Organization of the United Nations. Report no. B-13.

FAO. (2010b). Fish, crustaceans, molluscs, etc.; Capture production by species items; Africa-Inland waters. Food and Agriculture Organization of the United Nations. Report no. C-01(a).

FAO. (2010c). Tilapias and other cichlids; Capture production by species, fishing areas and countries or areas. Food and Agriculture Organization of the United Nations. Report no. B-12.

FAO. (2010d). Carps, barbels and other cyprinids; Capture production by species, fishing areas and countries or areas. Food and Agriculture Organization of the United Nations. Report no. B-11.

FAO. (2010e). Imports and exports by country and by seven fishery commodity groups. Food and Agriculture Organization of the United Nations. Report no. A-8.

FAO. (2012). Fishery and aquaculture country profiles: Malawi (December 8 2012; http://www.fao.org/fishery/countrysector/FI-CP_MW/3/en).

Fryer, G. (1960). Concerning the proposed introduction of Nile perch into Lake Victoria. *The East African Agricultural Journal*, 25: 267–270.

Gaden, M., Mkumbo, O.C., Lawrence, T. and Goddard, C. (2012). Top-down and bottom-up approaches in the management of the Laurentian Great Lakes and Lake Victoria fisheries: A comparison of two shared water bodies. pp. 364–390. *In*: Grover, V.I. and

Krantzberg, G. (eds.). Great Lakes Great Responsibilities: Lessons in Participatory Governance. Enfield, New Hampshire: Science Publishers.

Geheb, K. (1997). The regulators and the regulated: fisheries management, options and dynamics in Kenya's Lake Victoria fishery. Pages Socio-economic Data Working Group of the Lake Victoria Fisheries Research Project, Jinja. Lake Victoria Fisheries Project (LVFRP) - Phase II: LVFRP Technical Document No. 10. LVFRP/TECH/00/10.

Geheb, K. and Crean, K. (2003). Community-level access and control in the management of Lake Victoria's fisheries. *J. Environ. Manage*, 67: 99–106.

Goldschmidt, T. (1998). Darwin's Dreampond: Drama in Lake Victoria. MIT Press.

Hecky, R.E., Mugidde, R., Ramalal, P.S., Talbot, M.R. and Kling, G.W. (2010). Multiple stressors cause rapid ecosystem change in Lake Victoria. *Freshwater Biology*, 55: 19–42.

Kenyan Department of Fisheries. (2006). Implementation of a fisheries management plan for Lake Victoria in Kenya DoF, ed: Republic of Kenya.

Kudhongania, A.W. and Cordone, A.J. (1974). Batho-spatial distribution patterns and biomass estimate of the major demersal fishes in Lake Victoria. *African Journal of Tropical Hydrobiology and FIsheries*, 3: 15–31.

Kudhongania, A.W., Twongo, T. and Ogutu-Ohwayo, R. (1992). Impact of the Nile perch on the fisheries of Lake Victoria and Kyoga. *Hydrobiologia*, 232: 1–10.

Lawrence, T.J. (2015). Investigating the challenges and successes of community participation in the fishery co-management program on Lake Victoria, East Africa. Dissertation. University of Michigan.

LTA. (2012). Investment opportunities in Lake Tanganyika and Victoria are still limited by security and environmental threats (December 8 2012; http://lta.iwlearn.org/investment-opportunities-in-lake-tanganyika-and-victoria-are-still-limited-by-security-and-environmental-threats).

Lubovich, K. (2009). Cooperation and Competition: Managing Transboundary Water Resources in the Lake Victoria Region. FESS Working Paper No. 5. Foundation for Environmental Security and Sustainability: Falls Church, Virginia.

LVBC. (2017). Overview of LVBC (January 28 2017).

LVEMP. (2016). A brief of cage aquaculture in Lake Victoria, Kenya, 2016 Report.

LVFO. (2001). The convention for the establishment of the Lake Victoria Fisheries Organization: Lake Victoria Fisheries Organization in conjunction with the International Union for Conservation of Nature.

LVFO. (2005a). Guidelines for Beach Management Units (BMUs) on Lake Victoria. Jinja, Uganda: Lake Victoria Fisheries Organization, East African Community.

LVFO. (2005b). Book of abstracts: The state of the fisheries resources of Lake Victoria and their management: concerns, challenges and opportunities. Regional Stakeholders' Conference. Imperial Resort Beach Hotel, Entebbe, Uganda: Lake Victoria Fisheries Organization.

LVFO. (2011a). Welcome to Lake Victoria Fisheries Organization (Cited 2011 July 31 2011; http://www.lvfo.org).

LVFO. (2011b). Organs and institutions of the LVFO (December 4 2011; http://www.lvfo.org/index.php?option=com_content&view=article&id=52&Itemid=58).

LVFO. (2011c). Beach Management Units: Building co-management in East Africa (December 7 2011; http://www.lvfo.org/index.php?option=com_content&view=article&id=53&Itemid=59).

LVFO. (2015a). Fisheries Management Plan III (FMP III) for Lake Victoria Fisheries, 2016–2020. Lake Victoria Fisheries Organization.

LVFO. (2015b). Regional status report on Lake Victoria bi-ennial frame surveys between 2000 and 2014: Kenya, Tanzania and Uganda. Jinja, Uganda: Lake Victoria Fisheries Organization and East African Community.

Mkumbo, O.C. and Mlaponi, E. (2007). Impact of the baited hook fishery on the recovering endemic fish species in Lake Victoria. *Aquatic Ecosystem Health & Management*, 10: 458–466.

Mkumbo, O.C., Nsinda, P., Ezekiel, C.E., Cowx, I.G. and Aeron, M. (2007). Towards sustainable exploitation of Nile perch consequential to regulated fisheries in Lake Victoria. *Aquatic Ecosystem Health & Management*, 10: 449–457.

Naluwairo, R. (2005). A review of the national fisheries policy and the proposed fisheries legislation. Advocates Coalition for Development and Environment (ACODE). Report no. 22.

Njiru, M., Kazungu, J., Ngugi, C.C., Gichuki, J. and Muhoozi, L. (2008). An overview of the current status of Lake Victoria fishery: Opportunities, challenges and management strategies. *Lakes & Reservoirs: Research and Management*, 13: 1–12.

Njiru, M., Ngungi, P., Getabu, A., Wakwabi, E., Othina, A., Jembe, T. and Wekesa, S. (2006). Are fisheries management measures in Lake Victoria successful? The case of Nile perch and Nile tilapia fishery. *African Journal of Ecology*, 45: 315–323(319).

Njiru, M., Okeyo-Oworu, J.B., Muchiri, M., Cowx, I.G. and Van der Knaap, M. (2007). Changes in population characteristics and diet of Nile tilapia *Oreochromis niloticus* (L.) from Nyanza Gulf of Lake Victoria, Kenya: what are the management options. *Aquatic Ecosystem Health & Management*, 10: 434–442.

Njiru, M., Van der Knaap, M., Taabu-Munyaho, A., Nyamweya, C., Kayanda, B., Nyamweya, C.S., Kayanda, R.J. and Marshall, B.E. (2014). Management of Lake Victoria fishery: Are we looking for easy solutions? *Aquatic Ecosystem Health & Management*, 17: 70–79.

Njiru, M., Waithaka, E., Muchiri, M., Van der Knaap, M. and Cowx, I.G. (2005). Exotic introductions to the fishery of Lake Victoria: What are the management options? *Lakes & Reservoirs: Research and Management*, 10: 147–155.

North, D.C. (1990). Institutions, Institutional Change and Economic Performance. Cambridge University Press.

Nunan, F. (2010). Governance and fisheries co-management on Lake Victoria: Challenges to the adaptive governance approach. *Centre for Maritime Research (MAST)*, 9: 103–125.

Nunan, F., Cepic, D., Onyango, P., Yongo, E., Owii, M., Mlahagwa, E., Salehe, M., Odongkara, K. and Mbilingi, B. (2016). Networking for Fisheries Co-management on Lake Victoria, East Africa. Birmingham: International Development Department, University of Birmingham.

Nunan, F., Hara, M. and Onyango, P. (2015). Institutions and co-management in East African Inland and Malawi fisheries: A critical perspective. *World Development*, 70: 203–214.

Nunan, F., Luomba, J., Lwenya, C., Yongo, E., Odongkara, K. and Ntambi, B. (2012). Finding space for participation: fisherfolk mobility and co-management of Lake Victoria fisheries. *Environ. Manage.*, 50: 204–216.

Nunan, F. and Onyango, P. (2017). Inter-sectoral governance in inland fisheries: Lake Victoria. *In*: Song, A.M., Bower, S.D., Onyango, P., Cooke, S.J. and Chuenpagdee, R. (eds.). Inter-Sectoral Governance of Inland Fisheries. St. John's, NL, Canada: Too Big To Ignore.

Obiero, K.O., Abila, R.O., Njiru, M.J., Raburu, P.O., Achieng, A.O., Kundu, R., Ogello, E.O., Munguti, J.M. and Lawrence, T. (2015). The challenges of management: Recent experiences in implementing fisheries co-management in Lake Victoria, Kenya. *Lakes & Reservoirs: Research and Management*, 20: 139–154.

Odada, E.O., Ochola, W.O. and Olago, D.O. (2009). Drivers of ecosystem change and their impacts on human well-being in Lake Victoria basin. *African Journal of Ecology*, 47: 46–54.

Ogutu-Ohwayo, R. (1990). The decline of the native fishes of Lakes Victoria and Hyoga (East Africa) and the impact of introduced species, especially the Nile perch, Lates niloticus, and Nile tilapia, *Oreochromis niloticus*. *Environmental Biology of Fishes*, 27: 81–96.

Ogutu-Ohwayo, R. (2001). Efforts to incorporate biodiversity concerns in management of the fisheries of Lake Victoria, East Africa. Pages 1–20 in Programme UNDP-BPS, ed. Blue Millennium: Managing Global Fisheries for Biodiversity. Victoria, British Columbia: Global Environment Facility and United Nations Environment Programme.

Ojuok, J.E., Njiru, M., Ntiba, M. and Mavuti, K.M. (2007). The effects of overfishing on the life-history strategies of Nile talapia, *Oreochromis niloticus* (L.) in the Nyanza Gulf of Lake Victoria, Kenya. *Aquatic Ecosystem Health & Management*, 10: 443–448.

Ostrom, E. (2009). A general framework for analyzing sustainability of social-ecological systems. *Science*, 325: 419–422.
Pringle, R.M. (2005). The origins of the Nile perch in Lake Victoria. *BioScience*, 55: 780–787.
Republic of Kenya. (2008). National Oceans and Fisheries Policy 2008. Ministry of Fisheries Development.
Taylor, W.W., Bartley, D.M., Goddard, C.I., Leonard, N.J. and Welcomme, R. (eds.). (2016). Freshwater, fish, and the future: Proceedings of the global cross-sectoral conference Food and Agriculture Organization of the United Nations, Rome; Michigan State University, East Lansing; American Fisheries Society, Bethesda, Maryland.
Ugandan Department of Fisheries Resources. (2003). Guidelines for Beach Management Units in Uganda in Department of Fisheries Resources Uganda, ed. Ministry of Agriculture Animal Industry and Fisheries: Kampala, Uganda.
Ugandan Ministry of Agriculture Animal Industry and Fisheries. (2004). The National Fisheries Policy in Ministry of Agriculture Animal Industry and Fisheries, ed. Department of Fisheries Resources: Kampala, Uganda.
United Republic of Tanzania. (1977). The Constitution of the United Republic of Tanzania in Tanzania Go, ed. Duty to safeguard public property, Act No. 15 of 1984, Article 6: Dar es Salaam, Tanzania.
United Republic of Tanzania. (2011). The Tanzania five year development plan 2011/12-2015/16: Unleashing Tanzania's latent growth potentials. United Republic of Tanzania Presiden't Office, Planning Commission: Dar es Salaam, Tanzania.
Vaccaro, L. and Read, J. (2011). Vital to our nation's economy: Great Lakes jobs 2011 report. Ann Arbor, Michigan: Michigan Sea Grant.
Van der Knaap, M., Ntiba, M.J. and Cowx, I.G. (2002). Key elements of fisheries management on Lake Victoria. *Aquatic Ecosystem and Health Management*, 5: 245–254.
Van der Knaap, M. and Ligtvoet, W. (2010). Is western consumption of Nile perch from Lake Victoria sustainable? *Journal of Equatic Ecosystem Health Management*, 13: 429–436.
Van der Knaap, M. (2018). Are climate change impacts the cause of reduced fisheries production on the African great lakes? The lake Tanganyika case study. *In*: Johnson, J.E., De Young, C. and Virapat, C. (eds.). Proceedings of FishAdapt: Global Conference on Climate Change Adaptation for Fisheries and Aquaculture. Food and Agriculture Organization of the United Nations and Network of Aquaculture Centres in Asia-Pacific, FAO Fisheries & Aquaculture Proceedings No. x. Rome, Italy.
Wedig, K. and Stoehr, H. (in press). Water grabbing or sustainable development? Effects of aquaculture growth in neoliberal Uganda. *In*: Wiegratz, J., Martiniello, G. and Greco, E. (eds.). The Making of Neoliberal Uganda: The Political Economy of State and Capital after 1986. London: Zed Books.
Winfield, I.J., Winfield, Ian J., Baigún, Claudio, Balykin, Pavel A., Becker, Barbara, Chen, Yushun, Filipe, Ana F., Gerasimov, Yuri V., Godinho, Alexandre L., Hughes, Robert M., Koehn, John D., Kutsyn, Dmitry N., Mendoza-Portillo, Verónica, Oberdorff, Thierry, Orlov, Alexei M., Pedchenko, Andrey P., Pletterbauer, Florian, Prado, Ivo G., Rösch, Roland, Vatland and Shane, J. (2016). International perspectives on the effects of climate change on inland fisheries. *Fisheries*, 41: 399–405.
Witte, F., Wanink, J.H., Kishe-Machumu, M., Mkumbo, O.C., Goudswaard, P.C. and Seehausen, O. (2007). Differential decline and recovery of haplochromine trophic groups in the Mwanza Gulf of Lake Victoria. *Aquatic Ecosystem Health & Management*, 10: 416–433.
WWAP. (2012). The United Nations world Water Development Report 4: Managing water under uncertainty and risk. Paris, France: World Water Assessment Programme, United Nations Educational, Scientific and Cultural Organization.

CHAPTER 8

Lake Titicaca: Case Study

Paul Fericelli

Introduction

Lake Titicaca is the largest freshwater lake in South America and the highest of the large lakes in the world as it sits at an altitude of almost 4,000 meters above sea level. Waters from Lake Titicaca are evenly distributed between Bolivia and Peru, countries that agreed to follow a joint ownership model to ensure integrated management of the hydrological basin, which includes the Desaguadero River, and Lakes Poopó, Copaisa Salt and Titicaca (the TDPS System). The three main institutions that operate the TDPS System area: the Ministry of Sustainable Development and Planning, in Bolivia, the Peruvian Development Institute, in Peru, and the Binational Autonomous Authority of Lake Titicaca.

The environment surrounding the TDPS System is vulnerable to flooding and, increasingly, pollution. Extreme weather events, misuse of water supplies, uncontrolled economic activities, cultural differences in the region and the hydraulic complexity of the Lake Titicaca basin call for necessary policy and accountability reforms of national and binational water institutions governing the TDPS System. Currently, there are significant challenges in dispensing management responsibilities at different levels in water institutions resulting in uncoordinated efforts that aim to holistically use, protect and restore assets from the TDPS System.

This chapter presents a brief description of drivers impacting transboundary challenges in Lake Titicaca – environmental degradation,

---

Email: paul.fericelli@gmail.com

legal framework and climatic conditions – followed by an analysis of how conflicts in managing the region are resolved by Bolivia and Peru and their view on using dams in the TDPS System. Lastly, the chapter discusses how social/cultural differences influence restoration of Lake Titicaca.

Drivers Impacting Transboundary Challenges

The drivers impacting transboundary challenges in the TDPS System can be classified in three large groups: environmental degradation, legal framework and climatic conditions.

Environmental Degradation

Soil erosion, chemical, urban and industrial pollution are contributing to the environmental degradation of the TDPS System. Current erosion levels threaten agriculture in the Bolivian-Peruvian highlands altering the morphology and equilibrium of rivers due to the inflow of solid material that can impact water quality, especially when solid material, such as sediments, is chemically polluted. Chemical pollution in the TDPS System comes from acid mine drainage, which contaminates the water with lead, arsenic, and cadmium, and raw sewage discharges bringing high concentration of nutrients (World Water Assessment Programme, 2003, p. 474). The tropical weather areas in both countries and the high level of evaporation makes the Lake Titicaca watershed very vulnerable to pollution issues. Even though pollution levels may be maintained to arguably acceptable levels due to the size of Lake Titicaca, there are several cases of significant pollution near high-density communities surrounding it. For example, the eutrophication problems in Puno, Peru and the nutrient pollution in Cohana Bay, Bolivia.

Legal Framework

Legislation to conserve Lake Titicaca and benefit from its biodiversity have been promulgated by Bolivia and Peru since the early 90s, including laws, regulations and regional ordinances—some a result of international agreements. However, the laws, regulations and ordinances governing the TDPS System are partially applied due to the limited capacity and resources at different government levels in both countries (ALT, n.d.). The lack of effective application of the legal framework by government institutions exist due to circumstances, such as, lack of technical expertise, weak communication among state, municipal and district institutions, social resistance and rejection of positive ruling and application of vertical choice of law with no agreements among government institutions and stakeholders involved.

Climatic Conditions

The climatic conditions in the Lake Titicaca region vary largely and are prompt to extreme weather events. High levels of solar radiation, water diversion and cyclical El Niño droughts have led to extremely high evaporation rates to the point that Lake Poopó—a lake that is part of the TDPS System watershed and the second-largest lake in Bolivia—dried up in December 2015. Other extreme weather events in the TDPS System are related to flood risk conditions around Lake Titicaca significantly impacting the economy of the area for agriculture, animal raising and infrastructure. Hail and frost also have caused significant agricultural losses (World Water Assessment Programme, 2003, p. 477). All these events demonstrate that the TDPS System is not resilient to climate change affecting water availability for food, drinking water, sanitation and ecosystems survival.

Management Issues in Transboundary Context

Conflict Resolution

Bolivia and Peru have lead independent and joint efforts to prevent and resolve management conflicts in Lake Titicaca since 1957. Bolivia and Peru agreed to establish joint ownership where each country owns and manages half of the waters of Lake Titicaca. The countries conducted their own research concerning Lake Titicaca until the joint ownership agreement was ratified in 1986, then, established technical-oriented entities to coordinate actions with each national government and for centralizing information. In 1996, Bolivia and Peru approved the Binational Master Plan that was developed with the support of the European Community. The Binational Master Plan provides guidelines to control and preserve the TDPS System and a framework to soundly use its resources without hampering the environment.

The main government institution from each country that is in charge of designing, planning and enforcing policies, strategies and development initiatives for the TDPS System is: the Ministry of Sustainable Development and Planning, in Bolivia, and The Peruvian Development Institute, in Peru. Additionally, the Binational Autonomous Authority of Lake Titicaca (ALT) was established by both countries during the development of the Binational Master Plan to function as a joint autonomous management entity. ALT oversees the development of Lake Titicaca, mediates management conflicts through the Bolivian and Peruvian ministries of foreign affairs, provides a regulatory framework to both countries, and autonomously manages and implement actions for flood prevention, conservation and sound use of resources from the TDPS System.

ALT follows the integrated water resource management approach that promotes coordinated management and development of water, land and related resources without compromising the sustainability of vital ecosystems. Although ALT provides a regulatory framework to both countries, Bolivia and Peru have specific approaches for the management of Lake Titicaca.

In recognition of disparities in water management and to consolidate Lake Titicaca-related efforts, both countries signed a 10-year agreement on bilateral protective policies aimed to preserving Lake Titicaca in January 2016. The bilateral agreement focuses on reducing environmental pressures, implementing awareness activities, contributing to restore the lake's environment and biodiversity, and strengthening the understanding of the national comprehensive environmental management from each country. This agreement is the starting point to develop a management structure that facilitates strategic and holistic planning and reporting by setting common goals between the countries.

How are Dams Perceived in a Transboundary Context?

Bolivia and Peru perceive dams as necessary hydraulic regulation works that prevent, or at least protect, surrounding areas from floods in the TDPS System. The construction of the first dam in the TDPS System was completed in 2001 close to the international bridge over the Desaguadero River to protect fish populations and aquatic vegetation in the area. The Desaguadero Dam provides 50,000 hectares of secure irrigation to Peru, and 15,000 hectares yearly to Bolivia up to a maximum potential of 35,000 hectares, and flood protection for 6,000 to 10,000 hectares on both sides of the lake (World Water Assessment Programme 2003, p. 478). Even though the dam is one of the engineering solutions recommended by the Binational Master Plan for the needed hydraulic regulation works on both sides of the Desaguadero Lake, the operation of the dam has been questioned causing conflicts between authorities and stakeholders. For example, the scientific community from both countries claim that improper operation of the dam's gates is causing pollution and changes to ecosystems in the region affecting the economic and social development at the Desaguadero communities in Bolivia and Peru, and at the Poopó communities in Bolivia ("Compuerta provoca que el Titicaca se esté secando" 2014). In 2016, authorities and experts from the Poopó region in Bolivia, and abroad, claimed that improper operation of the Desaguadero dam was one of the causes of the Poopó Lake drying up (Howard 2016). ALT disagreed with the claims and stated that the drying up of the Lake Poopó was influenced by climatic conditions, needed dredging that has not been conducted in the entire Desaguadero region, sedimentation from mining activities and noncompliance with recommended actions described in the Binational Master Plan (Chuquimia 2016).

Science and Data Sharing

Under the 10-year agreement on bilateral protective policies, Peru and Bolivia agreed to design and implement a platform that facilitates management of binational information concerning environmental knowledge to register, disseminate and monitor recovery of the environment and biodiversity in Lake Titicaca – this, to have a baseline of such efforts from each country. Also, a binational monitoring system managed by homologous technical protocols applicable to both countries to measure water quality in Lake Titicaca will be established. Additionally, the countries agreed to prepare a binational agenda focus on environmental and biological diversity research including mechanisms for its implementation by institutions.

Cultural Differences

Differences in cultural patterns from indigenous and Western groups from Bolivia and Peru influence significantly the conservation and uses of the natural resources in the Lake Titicaca region, and are strongly limited and driven by a deteriorated environment and poverty (ALT, n.d.). The Lake Titicaca region encompasses diverse and autonomous indigenous groups with noticeable differences on cultural patterns that survived the colonization times and rejection from post-colonization societies. Among the cultural patterns from the Bolivian and Peruvian indigenous groups there are differences in languages, land ownership practices and economic activities practices, such as, agriculture, livestock raising and tourism. The differences in indigenous languages impact education and implementation of environmental conservation policies and efficient practices to improve economic activities. Even though the Aymara language is widely used in the entire Lake Titicaca, the Quechua language is important in the Peruvian side of the lake (ALT, n.d.). Hence, it is necessary to include both languages strategically in all community outreach efforts. Also, differences between indigenous and western cultural patterns make it difficult to adopt a new agricultural technology, improve production and establish efficient market systems (World Water Assessment Programme 2003, p. 477).

Conclusion

Bolivia and Peru have worked jointly under a cooperation framework to address the challenges of owning and managing Lake Titicaca; a transboundary lake with a growing environmental degradation, vulnerable to extreme climatic conditions and surrounded by poverty. Developing the Binational Master Plan with international input and establishing a joint autonomous institution has been very beneficial for the development of Lake Titicaca. For example, the establishment of the Binational Master Plan helped facilitate the process for obtaining funds for necessary scientific studies and joint infrastructure projects recommended in the plan. The Binational Master Plan provides valuable scientific information that help decision makers and recommends engineering solutions that aim for Lake Titicaca's sustainable development. However, Lake Titicaca is still being threatened by uncontrolled pollution and in need of restoration projects despite milestones reached by both countries today. This is due to tardy implementation of the recommended engineering solutions by the Binational Master Plan and failure to effectively implement the legal framework from each country at different government levels. Bolivia and Peru need to improve their current policies and rules that aim for the lake's preservation and have a stronger enforcement presence in the field. Both countries also need to include stakeholders and government institutions at municipal and district levels during the policy and rule making process to better educate, implement and enforce policies and rules.

The 10-year agreement signed by Bolivia and Peru in 2016 establishes the pillars of an improved integrated management framework for Lake Titicaca and milestones both countries have committed to reach to promote the lake sustainability. Also, the agreement establishes infrastructure projects to be financed by international stakeholders. Even though the agreement establishes concrete actions, Bolivia and Peru need to establish a management plan that also consider the national strategic planning and financial resources from each country.

Poverty is a major concern that exist in the Lake Titicaca region; the poor population in the region cannot afford to focus on environmental issues while striving to survive. Despite the political will, science-based objectives, and institutionalized cooperation in Bolivia and Peru, any polices enacted and actions to address water and environmental concerns need to have a poverty alleviation component to develop and improve living conditions in the region.

Beyond any doubt Bolivia and Peru have worked together aiming for sound management to develop Lake Titicaca in a sustainable manner and have had great accomplishments doing it, such as, the development of the Binational Master Plan, construction of the Desaguadero dam and the binational and international funding received historically. Also, both nations have demonstrated respect and cooperation while resolving conflict in managing Lake Titicaca. The most recent binational agreement outlines short-term and long-term milestones to be reached in 10 years. This will bring new opportunities for both nations to restore Lake Titicaca, reassess and improve current management practices at different government levels, and include accountability mechanisms to bring deterrence to actions that prevent the sustainable development of Lake Titicaca.

References

Autoridad Binacional Autónoma del Sistema Hídrico del Lago Titicaca, Río Desaguadero, Lago Poopó, Salar de Coipasa (ALT). (n.d.). Plan Estratégico de Conservación de la Biodiversidad del Sistema TDPS. Retrieve from ALT website: http://www.alt-perubolivia.org/Web_Bio/PROYECTO/planmaestro.html.

Bazoberry, A. (2015 June). Contaminación del Lago Titicaca, dragado del río desaguadero y represa. El Diario. Retrieved from: http://www.eldiario.net/noticias/2015/2015_06/nt150613/opinion.php?n=4&-contaminacion-del-lago-titicaca-dragado-del-rio-desaguadero-y-represa.

Chuquimia, L. (2016 January). ALT descarta que compuerta del Titicaca causen sequía del Poopó. Pagina Siete. Retrieved from: http://www.paginasiete.bo/sociedad/2016/1/10/descarta-compuertas-titicaca-causen-sequia-poopo-82914.html.

Comisión Binacional de Alto Nivel Perú – Bolivia. (2015). Lineamientos y Acciones para la Recuperación Ambiental del Lago Titicaca y su Biodiversidad Biológica. Retrieved from: http://www.minam.gob.pe/puno/wp-content/uploads/sites/55/2016/02/Documento-Lineamientos-y-Acciones.pdf.

Compuerta provoca que el Titicaca se esté secando. (2014 October). Correo. Retrieved from: https://diariocorreo.pe/ciudad/compuerta-provoca-que-el-titicaca-se-este-secando-471243/.

Howard, B. (2016 January). Bolivia's Second Largest Lake Has Dried Out. Can It Be Saved? National Geographic. Retrieved from: https://news.nationalgeographic.com/2016/01/160121-lake-poopo-bolivia-dried-out-el-nino-climate-change-water/.

MacQuarrie, Patrick, R., Welling, Rebecca and Mario Aguirre. (2013). Lake Titicaca Basin: Peru and Bolivia. Gland, Switzerland: IUCN. 12pp. Retrieved from: https://cmsdata.iucn.org/downloads/lake_titicaca_basin.pdf.

Ministerio del Ambiente Peru. (2016 February). Nuevo impulso para la sostenibilidad hídrica: GEF financiará proyecto de gestión integrada de los recursos hídricos en el sistema Titicaca-Desaguadero-Poopó-Salar de Coipasa [Press release]. Retrieved from: http://www.minam.gob.pe/puno/2016/02/23/nuevo-impulso-para-la-sostenibilidad-hidrica-gef-financiara-proyecto-de-gestion-integrada-de-los-recursos-hidricos-en-el-sistema-titicaca-desaguadero-poopo-salar-de-coipasa/#.

Newton, Joshua T. (2018). Case Study Transboundary Dispute Resolution: Lake Titicaca.

Revollo, M. (2010). Lake Titicaca: Experience and Lessons Learned Brief. International Waters Learning Exchange & Resource Network [Case Study].

United Nations World Water Assessment Program. (2003). The United Nations World Water Development Report: Water for People Water for Life. Retrieved from: http://unesdoc.unesco.org/images/0012/001297/129726e.pdf.

CHAPTER 9

Transboundary Governance in North America
More than 100 years of Development, Operation, and Evolution of the International Joint Commission

Gail Krantzberg* and Velma I. Grover

Introduction

Canada and the United States share the longest border in the world, of approximately 8,000 km, about 40% of which is shared water. Although most well-known are the Great Lakes, 300 lakes, rivers and streams define the Canadian–US border. Transboundary water resources can easily lead to upstream–downstream or 'tragedy of commons' types of problems (Grover and Krantzberg 2014).

The ownership and usage rights of these waters have been central to the industrial development of both countries, primarily providing easy waste disposal for the settlers of the early 20th century, particularly in the Great Lakes that lie directly between two heavily populated areas of these neighbouring countries (Wolf in Kenney 2005). At the turn of the 20th century, both nations formally recognized that a concerted binational

---

1280 main st. w., hamilton ontario, l8s 4k1, canada.
Email: velmaigrover@yahoo.com
* Corresponding author: krantz@mcmaster.ca

effort would be needed for the preservation of the trans-boundary water systems (deBoer and Krantzberg 2013). In 1909, the Boundary Waters Treaty was signed by both American and Canadian (British) Federal Governments. The Treaty states its purpose is to "provide the principles and mechanisms to help resolve disputes and to prevent future ones, primarily those concerning water quality and quantity along the boundary between Canada and the United States."

This chapter focuses on creation of Boundary Waters treaty (1909). A historical development of International Joint Commission (IJC) and evolution of the Great Lakes Water Quality Agreement (GLWQA) 1909 Boundary Waters Treaty (BWT) and evolution of IJC. The 1909 is a visionary treat that is one of the earliest innovations in transboundary governance which has later influenced many international governance and institutional developments (Muldoon 2012). A few reasons (among others) that led to the creation of BWT include: development of the regional without any comprehensive plan for hydropower development or water management issues, especially along Niagara and St. Mary's river; and no water management rules or regulations to resolve any of these issues and also no institution to developing any coordinated plans (Grover and Krantzberg 2014).

The conviction of those who negotiated the Boundary Waters Treaty was that solutions to the boundary problems should be based on deliberations of a permanent binational and equal institution, rather than through bilateral negotiations of diplomacy. The IJC promises equity without interfering with national sovereignty (Holmes 1981). The achievement of the common good as a basis for consensus has been the goal of the Commission for over 100 years (Krantzberg et al. 2006). Although the BWT and its related institutions have faced their own challenges, they have survived for over a century alongside the emergence of new treaties, agreements and institutions. In a way, the governance system has evolved over a period of time in response to the ecological, social and economic stresses (Muldoon 2012).

Through this treaty, the federal governments established the International Joint Commission whose role is/was to assist the two federal governments in finding solutions to the problems arising from the use of the trans-boundary water (IJC 2005). The IJC consists of six members; three appointed by the President of the United States and three appointed by the Governor in Council of Canada.

Structure and Operations of the IJC

The Commission has three principal functions:

1. Regulatory: It approves or disapproves applications from government, companies or individuals for obstructions, uses or diversion of water that can affect the natural level or flow of boundary water.[1]
2. Investigative: It investigates questions of difference, which are referred to the Commission by the two governments, and reports the facts to the two governments with recommendation for action. The governments decide whether or not to act upon the Commission's recommendations.
3. Surveillance/Coordination: It monitors compliance with the orders of approval. The IJC can monitor and coordinate actions or programs that result from the governmental acceptance of recommendations made to them by the Commission.

The Commission has six members, three on the part of the United States appointed by the President, and three on the part of Canada appointed by the Prime Minister on the recommendation of the Governor in Council. In practice, the IJC only becomes involved in assessing and advising on threats to the boundary region at the specific request of the federal governments; by way of the parties issuing the IJC a reference. Although the IJC may request a reference to address an issue it judges as worthy, there is no structure in place that allows for this to take place outside the direction of the two sovereign nations. As can be seen, according to the Treaty,

> "any questions or matters of difference arising between [the two countries] involving the rights, obligations, or interests of either in relation to the other or to the inhabitants of the other, along the common frontier between the United States and the Dominion of Canada, shall be referred from time to time to the International Joint Commission for examination and report, whenever either the Government of the United States or the Government of the Dominion of Canada shall request that such questions or matters of difference be so referred.
>
> The International Joint Commission is authorized in each case so referred to examine into and report upon the facts and circumstances of the particular questions and matters referred, together with such conclusions and recommendations as may be appropriate, subject, however, to any restrictions or exceptions

---

[1] This is of particular relevance when dams or obstructions are proposed by one party that could affect the other party to the Treaty.

which may be imposed with respect thereto by the terms of the reference."

The Commission is considered independent of the governments. It is not a supranational agency with legal authority. It theoretically operates on the principle of the reciprocal good of the resource. Each Commissioner upon the first meeting after being appointed signs a declaration to impartially perform the duties imposed under this treaty (Boundary Waters Treaty 1909). Krantzberg et al. (2006) point out that the Commissioners declare to seek the best solution to common problems based objectively on results of joint fact-finding studies. For technical information and policy advice in conducting investigations, the IJC depends mainly on boards or task forces with equal membership from each country. This facilitates binational consensus on science that is related to policy formulation, and enables data sharing among the members and both nations. Under the GLWQA, membership has been extended to nongovernmental experts, including representatives of environmental organizations and industry.

Evolution of Responsibilities of IJC and Creation of GLWQA

In 1964, Canada and the United States issued a highly influential reference to the IJC to study pollution in Lake Erie and elsewhere in the lower lakes that led to important changes in the Great Lakes regime. Scientists associated with the IJC found that excessive phosphorus loads from anthropogenic sources were resulting in severe eutrophication of Lake Erie and Lake Ontario. The 1964 reference eventually led to the creation of the Great Lakes Water Quality Agreement (GLWQA), one of the most significant contributions of the IJC to Great Lakes revitalization in its history (Krantzberg et al. 2006). The GLWQA is a permanent reference to the IJC and provides the Commission with duties beyond those in the BWT, most notably, to assist the Parties in implementing the Agreement and to assess progress in meeting the purpose of the Agreement.

The purpose of the 1972 Great Lakes Water Quality Agreement (GLWQA), is "to restore and maintain the chemical, physical and biological integrity of the waters of the Great Lakes Basin Ecosystem." Initially signed by then President Nixon and Prime Minister Trudeau, the GLWQA "set general and specific water quality objectives and mandated programs to meet them, however it gave priority to point source pollution from industrial sources and sewage plants" (IJC 2005). The GLWQA was successful in alleviating eutrophication (due to excessive nutrient inputs) and was revised in 1978 to better meet the emerging concerns over persistent toxic substances. Again in 1987, the GLWQA was amended, this time to include the development of Remedial Action Plans at geographic Areas of Concerns (AOC's) in the Great Lakes as well as Lakewide

Management Plans and specific commitments for nonpoint source pollution, contaminated sediment, airborne toxic substances, pollution from groundwater, and research and development priorities. Through successive revisions to the GLWQA, the governments have addressed emerging threats and harms, as they have become understood.

The GLWQA could be a unifying vision which aligns North American's protection and management strategies for the Great Lakes. For example, industrial, energy production, private consumption and recreational uses have different and often discordant requirements for water quantity and quality. Further, "mature issues" such as water pollution due to sewage and industrial by-products are reoccurring while new issues of emerging concern are continuing to arise, such as the presence and possible effects of pharmaceuticals and personal care products into the waste stream, air pollution and the ongoing introduction of new exotic species. This situation calls into question the nature of Great Lakes governance, institutional arrangements, and accountability, and highlights the need for reform of the Great Lakes regime.

The threats facing this globally significant fresh water resource are immediate. Consensus has been reached by hundreds of Great Lakes scientists, that the Great Lakes ecosystem's health is in jeopardy and facing a tipping point. Actions are urgently needed to restore system elements in critical near shore/tributary zones where a chain reaction of adaptive responses to a suite of stresses are leading to catastrophic changes – referred to as ecosystem meltdown. Without at least partial restoration of these areas, the adverse symptoms being observed in the Great Lakes will intensify and to a large extent become irreversible. Concurrently, actions are needed to control or eliminate sources of stress that represent basin-wide threats to the biological, physical and chemical components of the Great Lakes essential to stability and health of the ecosystem (Beeton et al. 2005).

The IJC's Twelfth Biennial Report called for the renewal of GLWQA, since at that time it had not been updated in more than 17 years while science and technology had grown substantially. As such, "we need to keep pace with what we know and review the agreement with an eye toward a sustainable future" (IJC 2004, Jetoo and Krantzberg 2014a). The governments finally started the review process in 2006 and the revised protocol (after long public consultations and comments) was signed in 2012. The purpose of the 1987 GLWQA is still part of the new protocol, but its scope now includes current issues such as climate change, aquatic alien invasive species, habitat and species, and near-lake shore areas (IJC 2011) of the Great Lakes Basin. IJC (2011) identified governance as a key issue in its 15th Biennial Report, noting that "there is a critical need to modify existing governance to strengthen coordination across jurisdictional lines to address ecological challenges in the nearshore" (Jetoo and Krantzberg 2014b). Another key inclusion is the engagement of First Nations, tribal governments, Metis and municipal governments. The GLWQA Protocol of 2012 heeded the calls of the public and the IJC to incorporate previously unaddressed issues such as climate change with the inclusion of three new annexes; climate change, habitat and species and aquatic invasive species. It was recognized in the introduction that the Protocol is placing emphasis on addressing new and emerging threats to the waters of the Great Lakes.

One of the strengths of the Protocol is in the clear depicting of the role of the IJC, which retains its oversight, public information and investigative roles. Article 7(k) describes the triennial reporting requirement utilizing the IJC Advisory Boards, to the Parties to the agreement. Article 8(3) and (4) clarify the roles of the WQB and the Science Advisory Boards (SAB). The WQB is the principle policy advisor to the IJC assessing progress of the Parties while the SAB will provide advice on science and research matters. The shift from biennial to triennial reporting will allow the IJC time to gather and assess data and provide a more comprehensive report. These changes will likely be welcomed by the Great Lakes Community who attributed the lack of comprehensive data reporting that failed to document the true state of the Great Lakes ecosystem since the early 1990s as a direct result of the curtailed function of the IJC (CELA 2006).

Another strength of the 2012 protocol reflects concerns over climate change impacts on the integrity of the basin waters. Annex 9 of the GLWQA of 2012, Climate Change Management, calls for both Parties, in collaboration with other orders of government and nongovernment actors, to:

1. "develop and improve regional scale climate models to predict climate change in the Great Lakes Basin Ecosystem at appropriate temporal and spatial scales;
2. link the projected climate change outputs from the regional models to chemical, physical, biological models that are specific to the Great Lakes to better understand and predict the climate change impacts on the quality of the Waters of the Great Lakes;
3. enhance monitoring of relevant climate and Great Lakes variables to validate model predictions and to understand current climate change impacts;
4. develop and improve analytical tools to understand and predict the impacts, and risks to, and the vulnerabilities of, the quality of the Waters of the Great Lakes from anticipated climate change impacts; and
5. coordinate binational climate change science activities (including monitoring, modeling and analysis) to quantify, understand, and

share information that Great Lakes resource managers need to address climate change impacts on the quality of the Waters of the Great Lakes and to achieve the objectives of this Agreement" (GLWQA 2012).

Legal Mechanisms to Incorporate the Protocol

Both the United States and Canada have existing legal mechanisms that enshrine parts of the protocol in cooperative agreements and law. For example, Canada has relied on the Canadian–Ontario Agreement Respecting the Great Lakes Basin Ecosystem (COA) as a mechanism for cooperation between the Province of Ontario and the federal government for Great Lakes Restoration. The Ontario Ministry of Environment and Climate Change led to the establishment of Great Lakes Protection Act[2] which has the potential to provide tools for setting broad direction for ecological restoration as well as accommodating targeted action in priority degraded areas. Similarly the United States has recognized the GLWQA in the Clean Water Act, the Beaches Act, the Great Lakes Restoration Initiative (GLRI) and several Executive Orders of its presidents. These are visionary precedents that can aid in the implementation of the protocol (Jetoo and Krantzberg 2014b).

Characteristics of the Canada–US Transboundary Water Governance Inter-regime

This section analyzes the IJC related regime based on five elements of public governance that are helpful for analyzing the IJC-related regime (deBoer and Krantzberg 2013). These elements are the scales, actors, goals, instruments and resources (Bressers and Kuks 2003).

Actors are seen to operate in multi-level and multi-actor networks. Goals result from multiple problems. The means available are both instruments and resources for their implementation and are characterized by multiple instruments that are joined together in strategic policy mixes. All parties involved use formal and informal resources in interactive implementation. This set of elements uncover critical characteristics that could enable more sustainable water governance.

Elements of public governance can be used to describe the aspects of the governance inter-regime that affect the ability of the goals to be achieved. When looking at the governance inter-regime as a whole, it is further helpful to assess how each of these elements contribute to four essential regime qualities as are expressed within Contextual Interaction

---

[2] https://www.ontario.ca/laws/statute/15g24.

Theory: extent, coherence, flexibility and intensity (De Boer and Bressers 2011).

In the early stages, most simple governance regimes have little extent, i.e., they begin with some piece of legislation, rule or law that attempts to reduce some negative impact on the resource. As more attention gets paid to the impact of different uses and activities, the extent of the regime increases by way of additional instruments to protect the resource (Knoepfel et al. 2001, 2007). As this happens, the number and breadth of instruments, actors, and influences increase and generally results in decreasing coherence since more complexity leads to fragmentation. In a more integrated inter-regime, these incoherencies are addressed through various attempts to either streamline or better coordinate the various pieces of the different regimes, which influence the resource. It is postured in Bressers and Kuks (2004) that as the regime moves from low extent and coherence, towards higher extent and coherence that the resulting regime (likely to resemble our definition of inter-regime) is more apt to ensure more sustainable management of the given resource. Lacking extent leaves important uses and users unregulated and lacking coherence leads to measures that hinder each other's effectiveness, which naturally can also be seen as a lack of connective capacity of the institutional structure.

In a more dynamic adaptive management perspective two additional criteria are relevant for understanding how the governance inter-regime supports the sustainable management of the resource (De Boer and Bressers 2011). Like with economic development, adaptive action in complex and dynamic circumstances requires many actors' creativity and effort. It is important that the governance inter-regime rewards initiatives from many stakeholders as long as they contribute to the common vision and also provides leeway for such initiatives.

The first quality added to the framework is the aspect of *intensity*. This is the degree to which the inter-regime elements urge changes in the status quo or in current developments. Secondly the *flexibility* of the inter-regime characterizes the degree to which the governance inter-regime elements support and facilitate adaptive actions and strategies (in as far as the integrated ambitions are served by this adaptiveness, not as a result of weak implementation. Increased flexibility is likely to support the sustainable management of the resource when the cognitions and the motivations of the various implementers are in line with those of the policy developers and decision makers. Increasing the flexibility of an inter-regime beyond a certain point however is unlikely to further increase sustainability and even leads to a reduction due to a lack of clarity and responsibility for action or accountability. Accordingly, too little or too much intensity in an inter-regime under the non-corresponding conditions can lead to inappropriate levels of forced implementation which is generally conceptualized as not supporting the trust and social sustainability of the system (deBoers and Krantzberg 2013).

Great Lakes Transboundary Inter-Regime Context

Extent

The extent of the inter-regime appears at first glance to be very high due to the numerous and extensive levels and scales, actors, problem perceptions and objectives, instruments and policy strategies included within it. What is interesting is that most concerns about extent were generally related to the idea that there are so many policies and programs dealing with the Great Lakes that are relatively unclear, that it is too difficult to identify where the actual gaps are in terms of governance of the resource. Resources are not available at the local group level to support the place-based action. This lack of resources and general lack of clarity reduce the ability of the inter-regime to support adaptive water management.

Coherence

There are a large number of government levels, institutions, NGOs and the public involved in the governance inter-regime. This is an extremely important characteristic of this inter-regime. It has developed partially due to the understanding that it is a shared resource both binationally and locally and thus the more people that are involved, the less likely it is that one of those bodies will take from the resource in an unsustainable way. This can have other than the intended results. For purposes of clarity, it is instructive to understand how these numerous instruments can work against the sustainable management of the resource based on their overall coherence.

Logistically, there appear to be just too many levels and scales involved for effective negotiations to take place when all the parties who have a stake are involved since the issue itself is considered to be so large. Increasing decision-making at the local level over aspects important to the success of transboundary water governance has also reduced the faith in proper accountability and clear progress. A large number of the actors in the network have similar interests, although this is certainly not the case across the board. A general difficulty in bringing the actors together in consensual network relations is certainly a strain on the coherence of governance.

GLWQA and the surrounding inter-regime both work toward putting the interests of the lakes on the national agenda of each country. Policy strategies at the national levels in the US however, are not supposed to be heavily influenced through regional issues. This diminishes the ability for the EPA to develop specific policies related to the Great Lakes.

US and Canadian policy instruments and strategies differ and thus make joint agreements on measures more difficult (note that this is especially difficult where increased regulative detail is sought as opposed to focusing on a guiding vision, the achievement of results and the use flexibility to accomplish them). In addition, the manner of appointing the IJC is not delivering confidence in its impartiality. Concerns are strong that the recent politicizing of the IJC appointments has reduced the ability of the IJC to effectively act as an instrument to manage the lakes binationally. There are also constraints to integrating binational decisions into domestic law on both sides of the border.

The incoherencies continue further down the implementation trail. One example is that municipal governments have the responsibility to enact the GLWQA but are not given the resources to do so, and are requested to fit these actions into their agendas.

Flexibility

An inter-regime is more flexible when there is a decentralization of power that is supported by upper levels of government. This is closely related to empowering rather than controlling relations, and thus relies on trust. The high stakes that the two federal governments' hold for the control of these shared waters is certainly seen to inhibit the reform of the inter-regime towards supporting solutions that can come from local adaptive actions. The transfer of priority setting and assessment of governments' progress to an impartial third body such as the IJC would reduce such control, yet that is unlikely to change. Thus, the lack of trust across levels and the desire to hold decision-making power at the upper levels reduces the flexibility of the inter-regime for local stakeholders and thus hinders their ability to act closely in relation to addressing local contextual issues.

The strategies and instruments are more flexible when means from different sources (like public policies and private property rights) are available and can be used in creative ways to achieve the desired goals. This aspect is difficult to assess, as there are indeed numerous varieties and types of instruments available as a result of the broad issue areas that are related to the Great Lakes. What makes it problematic is that in this context, this flexibility enables the incoherencies to be further exploited in ways that are perhaps more difficult to coordinate along shared understanding of the intent of the inter-regime. Thus, it could be said that this flexibility is quite high, but this does not necessarily contribute to the ability of the inter-regime to sustainably manage the resource.

Generally it can be said that flexibility is usually discouraged from the top level due to a strong history of using command and control methods to ensure that the implementing agencies are accountable to the interests of the government. This is as well the case from the lower levels due to a lack of trust in the overall regime. It is however clear that increased flexibility could support a more coherent and well-functioning overall regime if it were aided by a significant increase in the presence of trust and alignment of cognitions and motivations between the various levels of policy and implementation.

Intensity

The scale as well as the dynamic and change-oriented nature of the Great Lakes inter-regime requires energy to move forward and overcome current obstacles. Political will to act in the binational interest was seen as very weak against the lobbying interests of specific interest groups and sectors related mostly to industrial and economic desires. The general strength and influence of international law is important in underpinning some of the perceived lack of intensity of the inter-regime. There is certainly no consensus with respect to the "binding" nature of the Great Lakes Water Quality Agreement as the political will of the day is seen to be what gives the various instruments real importance and chances for successful implementation. The funding connection between the IJC and the federal governments alters the ability of the IJC to be impartial and hence force difficult decisions in the direction where they are more politically safe. The responsibility for enforcement of the Agreement is unclear and so the lack of desire at the federal level to enforce environmental obligations and provide the necessary resources for implementation (perhaps as a result of incoherent problem definitions and objectives) causes this to significantly reduce the intensity of the inter-regime in protecting the lakes.

Given the dwindling capacity that has taken place in the grass roots community over the past decades, these reductions in intensity at the inter-regime level are even more dangerous to the sustainability of the lakes. Botts and Muldoon (2005) express the concern that the governments of the Great Lakes are not sufficiently mobilized for coherent action and that there needs to be a revitalization of the binational sense of community amongst these neighbours for the future good of the Great Lakes. One example that is cited of the breakdown and distrust in the current decision and policy making structure is the withdrawal of the USEPA and Environment Canada from the institutional arrangements managed by the IJC in the 1990s. Although there are demands for the commission to take on existing and new challenges, Lemarquand (1993) shows that "the IJC operates best within a fairly narrow range of issues, including boundary water project supervision, fact finding, and evaluation. Going beyond this range would likely result in failure. Nevertheless, as boundary environmental relations change from conflictual to more managerial, the IJC could productively take on an expanded role in evaluation and the provision of necessary information. The governments, however, are failing to make use of this potential". Allee (1993) concurs that the role of the IJC has been that of a local authority on given issues, an institution for the federal governments to consult after a conflict has been reduced to a technical issue or where the IJC's study role can serve as a step toward achieving that political result.

Public participation has been central to the success of IJC, GLWQA and RAPs program. As explained by Hall (2009), "Since the 1970s, the second generation (after the BWT) of environmental agreements between the United States and Canada demonstrate a dramatic growth in the role of citizens in achieving compliance with international environmental law."

The GLWQA relies substantively on citizens to press for compliance of the Parties with their commitment, and recognize that the two federal governments may have more in common with each other than with citizens and other stakeholders on both sides of the border regarding environmental protection and injury. The IJC cannot accomplish bilateral cooperation under the Treaty or achieve the objectives of ecosystem rehabilitation and protection under the Great Lakes Water Quality Agreement without the support and concerted action of a well-informed and educated citizenry, including a well organized and effective binational nongovernmental organizational presence to hold the governments accountable (Becker 1993).

The IJC's strengths as a third party adviser, a fact finder and technical mediator, an environmental assessment project evaluator, an overseer keeping certain types of issues off the bilateral agenda, a consensus builder, and increasingly as an evaluator under the Great Lakes Water Quality Agreement are the types of qualities required in the more depoliticized and managerial setting that may be emerging. Lemarquand's finding in 1993 holds true in 2018. The most promising areas of reform would place less emphasis on the historical strengthening of institutional effectiveness and efficiency in arriving at technical recommendations. Rather, reform would move the IJC further into the area of generating information and knowledge in response to public and government needs, and in advancing objective evaluations that push governments to be more politically accountable to their citizens in demonstrating progress in dealing with bilateral environmental issues.

Over the last several decades, in the Great Lakes basin, and elsewhere, there has been a transition from, for example, water quality and fisheries management regime into the new ecosystem management paradigm. The shift has affected both scientific and management thinking and has brought many more agencies and stakeholders into the fray. The emergence of the ecosystem approach was in part a response to the inadequacy of the water quality approach. It has not, however, brought with it a shift in governance mechanisms (Krantzberg 2007, Krantzberg et al. 2015). If the GLWQA is to encompass habitat and biodiversity, and climate change adaptation, for example, the governing responsibilities and relationships will need to reflect this.

Optimism remains that the Great Lakes will be revitalized under the 2012 GLWQA protocol. Indigenous shareholders, and citizens from governments, the private sector, and civil society need to share a vision of shared, effective, inclusive and responsible governance to protecting and enhancing this global treasure. With this, we can make the Lakes Great.

Conclusion

Effective transboundary lake basin governance is founded on a shared desire to respond to environmental, socio-economic, legal, and management challenges and help deal with conflict. Transboundary lakes, by definition, cross national boundaries; consequently, management and use of natural resources are subject to the policies and laws of more than one political jurisdiction (UNU-INWEH 2011). Because resources are shared, actions in one jurisdiction can affect others. Effective management requires cooperation and harmonization of water and environmental management policies, predicated on collaborations on science and technology. Transboundary commissions provide an excellent forum for planning, but responsibility to implement programs generally rests with governments and other cooperators. In the Laurentian Great Lakes, implementing plans works best when there is both "top-down and bottom-up" connectivity and communication among commissioners, high-level government officials and resources managers, and the technical staff responsible for the "on-the-ground" research. Further, the IJC has a long history of public participation in their activities, which is often instrumental in assuring accountability and translating public pressure into political will for implementing priority transboundary strategies (UNU-INWEH 2011).

Clearly, the manner in which Great Lakes governance challenges are diagnosed and addressed is of interest to those seeking to protect the sustainability of transboundary waters in other regions in the face of similar challenges. What is needed, state VanNijnatten et al. (2016), is a deeper understanding of the conditions which promote effective transboundary governance, in order that we might identify the appropriate means to tackle these "wicked problems" and fill capacity gaps in current governance frameworks. There are a range of attributes associated with measuring governance capacity, including high levels of leadership; necessary and sufficient participation; shared discourse and mutual understanding; sustainable resources; and a strong institutional basis. The presence of a strong institutional basis, while not necessarily *more* important than the other attributes, is foundational in terms of Governance Capacity, as institutions act to channel policy discourse, structure policy choices and resources, and provide opportunities or constraints for policy actors.

VanNijnatten et al. (2016) demonstrate that the primary challenge, according to every major international body studying water, is one of "governance"; as the Organization for Economic Cooperation and Development [OECD] (2011) has stated: "the current water 'crisis' is not a crisis of scarcity but a crisis of mismanagement, with strong public governance features." Similarly, Biswas and Tortajada (2010) argue that it is not the "physical scarcity of water" that is the major problem, but rather "the past and current poor to very poor governance practices . . . used in nearly all developing and developed countries." The proliferation of academic and policy work on water governance highlights the continuing search for effective institutions and processes for creating and maintaining sustainable water regimes.

References

Allee, D.J. (1993). Subnational governance and the international joint commission: Local management of United States and Canadian Boundary Waters. *Natural Resources Journal*, 33: 133–151.

Becker, M.L. (1993). The international joint commission and public participation: Past experiences, present challenges, future tasks. *Natural Resources Journal*, 33: 235–274.

Beeton, R., Henry Regier, Jack Bails, Dale Bryson, Jonathon Bulkley, Wil Cwikiel, John Gannon and Mike Murray. (2005). Prescription for the Great Lakes: www.great-lakes.net/sdstrategyteam/documents/Prescription_for_Great_Lakes_3-28-05.doc.

Biswas, A.K. and Tortajada, C. (2010). Future water governance: Problems and perspectives. *Water Resources Development*, 26(2): 129–139.

Botts, Lee and Paul Muldoon. (2005). Evolution of the Great Lakes Water Quality Agreement. Michigan State University Press, 2005. JSTOR, www.jstor.org/stable/10.14321/j.ctt163t7st.

Boundary Waters Treaty. (1909). Treaty Between The United States and Great Britain Relating to Boundary Waters, and Questions Arising Between The United States and Canada. http://ijc.org/files/tinymce/uploaded/Boundary%20Waters%20Treaty%20of%201909_3.pdf.

Bressers, H.Th.A. and Kuks, S. (2003). What does "Governance" mean? From conception to elaboration. pp. 65–88. *In*: Bressers, H.Th.A. and Rosenbaum, W.A. (eds.). Achieving Sustainable Development: The Challenge of Governance Across Social Scales. Westport, CT: Praeger.

Bressers, H.Th.A. and Kuks, S. (2004). Integrated Governance and Water Basin Management. Dordrecht, Boston, London: Kluwer Academic Publishers.

Canadian Environmental Law Association (CELA). (2006). The Great Lakes Water Quality Agreement; Promises to keep, challenges to meet. In consultation with Great Lakes Environmental Community. CELA, Great Lakes United, Alliance for the Great lakes, Biodiversity Project.

De Boer, C. and Bressers, H.Th.A. (2011). Complex and Dynamic Implementation Processes: The Renaturalization of the Dutch Regge River. Enschede: University of Twente Press.

deBoers, C. and Krantzberg, G. (2013). Great lakes water governance: A transboundary inter-regime analysis. pp. 313–332. *In*: Jurian Edelenbos, Nanny Bressers and Peter Scholten (eds.). Water Governance as Connective Capacity (Ashgate).

Grover, V.I. and Krantzberg, G. (2014). Transboundary water management: lessons learnt from North America. Water International. http://dx.doi.org/10.1080/02508060.2014.9 84962.
Hall, N.D. (2009). Introduction: Canada–United States transboundary environmental protection. *Windsor Review of Legal and Social Issues*, 26: 1–6.
Holmes, J.W. (1981). Introduction to the international joint commission seventy years on. pp. 3–7. *In*: Spencer, R., Kirton, J. and Nossal, K.R. (eds.). Centre for International Studies, University of Toronto ISBN 7727-0801-4.
IJC [International Joint Commission]. (2004). Twelfth Biennial Report.
International Joint Commission. (2005). A Guide to the Great Lakes Water Quality Agreement: Background for the 2006 Governmental Review ISBN 1-894280-53-9.
IJC [International Joint Commission]. (2011). Fifteenth Biennial Report.
Jetoo, S. and Krantzberg, G. (2014a). Donning our thinking hats for the development of the Great Lakes nearshore governance framework. *Journal of Great Lakes Research*, 40: 463–465. doi:10.1016/j.jglr.2014.01.015.
Jetoo, S. and Krantzberg, G. (2014b). A SWOT analysis of the Great Lakes Water Quality Protocol 2012: The good, the bad and the opportunity. *Electronic Green Journal* (http://escholarship.org/uc/item/7h26v4cv).
Knoepfel, P., Kissling-Näf, I. and Varone, F. (2001). Institutionelle Regime für natürliche Ressourcen. Basel: Helbing & Lichtenhahn.
Knoepfel, P., Nahrath, S. and Varone, F. (2007). Institutional regimes for natural resources: an innovative theoretical framework for sustainability. pp. 455–506. *In*: Knoepfel, P. (ed.). Environmental Policy Analyses: Learning from the Past for the Future: 25 Years of Research. Berlin: Springer.
Krantzberg, G. (2007). The Great Lake's future at a cross road. *The Environmentalist*. DOI 10.1007/s10669-007-9139-z.
Krantzberg, G., Bratzel, M. and McDonald, J. (2006). Contribution of the international joint commission to great lakes renewal. *The Great Lakes Geographer*, 13: 25–37.
Krantzberg, G., Irena, F. Creed, Kathryn, B. Friedman, Katrina, L. Laurent, John, A. Jackson, Joel, Brammeier and Donald Scavia. (2015). Community engagement is critical to achieve a "thriving and prosperous" future for the Great Lakes–St. Lawrence River basin. *Journal of Great Lakes Research*, 41 Supplement 1: 188–191.
Lemarquand, D. (1993). The international joint commission and changing Canada-United States boundary relations. *Natural Resources Journal*, 33: 60–91.
Muldoon, P. (2012). Governance in the Great Lakes – A regime in transition. *In*: Grover, V.I. and Krantzberg, G. (eds.). Great Lakes: Lessons in Participatory Governance. Boca Raton: CRC Press, Taylor and Francis Group.
Organization for Economic Cooperation and Development. (2011). Water governance in OECD countries: A multi-level approach. OECD Studies on Water. Author. Retrieved from http://www.waterforum.net/Content/Artikel/4000-5000/4243/OECD_rapport_water_governance_130220104537.pdf.
UNU-INWEH. (2011). Transboundary Lake Basin Management: Laurentian and African Great Lakes, UNU-INWEH, Hamilton, Ontario, Canada, 46 p.
VanNijnatten, Debora, Carolyn Johns, Kathryn Bryk Friedman and Gail Krantzberg. (2016). Assessing adaptive transboundary governance capacity in the Great Lakes Basin: The role of institutions and networks. *International Journal of Water Governance*, 1: 7–32.
Wolf, A.T. (2005). Transboundary water conflicts and cooperation. pp. 131–54. *In*: Kenney, D.S. (ed.). Search of Sustainable Water Management. Northampton, MA: Edward Elgar.

Index

A

accountability mechanisms 157
Area of Concerns 74, 161

B

bilateral protective policies 154, 155
Binational Master Plan 153–156
Boards 161, 163
BWT 159, 161, 169

C

Climate change 7, 8, 67, 71, 75, 162–164, 169
community participation 131

D

Drin River Basin 78, 94, 97, 102, 103, 106, 107

F

fisheries 130–144, 146

G

GLRI 74, 75, 164
GLWQA 1, 5, 6, 71, 72, 159, 161–164, 166, 167, 169, 170
Governance 11–33, 111–113, 115, 117, 118, 121–123, 127, 130, 131, 133, 134, 138, 140–143, 146
ground water 3, 40–53, 55, 58–60
groundwater governance 40–45, 47–49, 55, 58–60

H

HELCOM 5

I

IJC 70, 71, 159–164, 167–170
indicators 11, 13, 14, 17–28, 30–33

integrated water resources management 154
international cooperation 45
Invasive Species 7

J

joint autonomous institution 156
joint ownership 151, 153

L

Lake Titicaca 6, 7, 151–157
Lake Victoria 130–137, 139–146

M

management 131–143, 145, 146
Mexico 42, 46, 53–56, 58, 60

O

Ohrid Lake 83, 89–92, 94, 97, 98, 102, 107

P

Prespa Lakes 83, 89–92, 94, 97, 98, 102, 107

R

Remedial Action Plans 161

S

Science Advisory Boards 163
Skadar Lake 83, 98, 101
stakeholder engagement 8

T

The Baltic Sea 111–113, 116, 117, 120, 122–124, 126
The ecosystem approach 111–113
The North American Great Lakes 111
transboundary cooperation 132
Transboundary Water Commissions 111, 112

Transboundary Water Governance 80
transboundary waters 52, 56, 60

U

United States 42, 53, 55, 56

W

water governance indicators (WGI) 3, 4
water policy 12, 15, 19, 20, 22, 25
Water Quality 159, 161, 168, 169
water resources agreement 72–74
WQB (Water Quality Boards) 163

PGSTL 08/15/2018